AA002338

2012 IEEE Workshop on Microelectronics and Electron Devices

(WMED 2012)

Boise, Idaho, USA
20 April 2012

IEEE Catalog Number: CFP12564-PRT
ISBN: 978-1-4577-1735-2

**Copyright © 2012 by the Institute of Electrical and Electronic Engineers, Inc
All Rights Reserved**

Copyright and Reprint Permissions: Abstracting is permitted with credit to the source. Libraries are permitted to photocopy beyond the limit of U.S. copyright law for private use of patrons those articles in this volume that carry a code at the bottom of the first page, provided the per-copy fee indicated in the code is paid through Copyright Clearance Center, 222 Rosewood Drive, Danvers, MA 01923.

For other copying, reprint or republication permission, write to IEEE Copyrights Manager, IEEE Service Center, 445 Hoes Lane, Piscataway, NJ 08854. All rights reserved.

***This publication is a representation of what appears in the IEEE Digital Libraries. Some format issues inherent in the e-media version may also appear in this print version.**

IEEE Catalog Number: CFP12564-PRT
ISBN 13:   978-1-4577-1735-2
ISSN: 1947-3834

Additional Copies of This Publication Are Available From:

Curran Associates, Inc
57 Morehouse Lane
Red Hook, NY 12571 USA
Phone: (845) 758-0400
Fax: (845) 758-2633
E-mail: curran@proceedings.com
Web: www.proceedings.com

2012 IEEE Workshop on Microelectronics and Electron Devices (WMED 2012)

Boise, Idaho, USA
20 April 2012

IEEE Catalog Number: CFP12564-POD
ISBN: 978-1-45771-735-2

2012 IEEE Workshop on Microelectronics and Electron Devices (WMED)

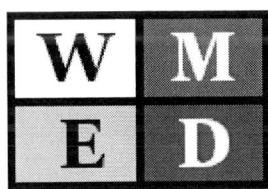

Jordan Ballroom, Student Union Building
Boise State University

April 20, 2012

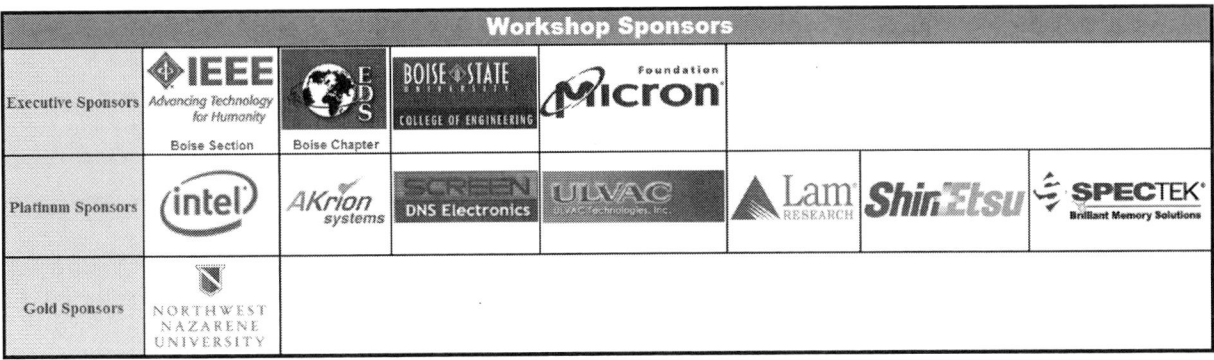

This workshop is receiving technical co-sponsorship support from the IEEE Electron Devices Society.

Welcome to the 2012 IEEE WMED

Dear Participant,

It is my pleasure to welcome you to the 2012 IEEE Workshop on Microelectronics and Electron Devices (WMED). The WMED is an IEEE professional society workshop hosted by the Boise Chapter of the IEEE Electron Devices Society (EDS). The WMED brings together students, faculty and industry researchers for a day of engaging invited talks and tutorials, technical presentations, poster sessions, and professional networking. This year marks our tenth anniversary of hosting the WMED in Boise. We believe that, over the last nine years, the workshop has played a key role in promoting the sharing of technical knowledge and information in the Northwest region of the United States and beyond. I am thrilled to see a strong interest in our workshop again this year. The 2012 WMED management committee and I believe that this year's program will be interesting and educational for all of you.

We are very excited to have several distinguished speakers from academia and industry who will cover a wide range of topics in the semiconductor field. I would like to thank our invited speakers and contributors for sharing their knowledge and expertise with us today.

During the workshop, IEEE WMED also provides a unique program for local area high school students. The WMED high school program introduces students to the field of engineering as a career choice, and features presentations highlighting engineering experiences from experts in the field, through technical talks and a panel discussion.

I would like to thank each of our sponsors for their generous financial support. This workshop would not have been possible without the contributions of all the members of the 2012 WMED management committee as well as the volunteers. I am also grateful to the WMED Advisors and the University Advisory Board for their helpful suggestions and support. Finally, I would like to thank all the participants of this workshop for making this event possible.

Best Regards,

Prashant Raghu

General Chair, IEEE WMED 2012

IEEE WMED 2012 Management Committee

Prashant Raghu – General Chair
Tim Hollis – Technical Chair
Jaydeb Goswami – Technical Vice Chair
Suraj Mathew – Publications Chair
Jaydip Guha – Program Chair
Roland Awusie – Program Vice Chair
Sanjeev Sapra – Registration Chair
Huy Le – Finance Chair
Sourabh Dhir – Finance Vice Chair
Eric Booth – Publicity Chair
Roy Meade – Publicity Vice Chair
Randy Wolff – High School Program Chair
Laurie Anderson – High School Program Advisor
Shyam Surthi and Vishwanath Bhat – Senior Advisors
Archana Nanjundarao – Webmaster
Hari Naidu – Photographer

IEEE WMED 2012 University Advisory Board

Prof. Maria Mitkova, Boise State University
Prof. Vishal Saxena, Boise State University
Prof. Stephen Parke, Northwest Nazarene University

IEEE Electron Devices Society (EDS) Boise Chapter Officers

Shyam Surthi – Chapter Chair and Secretary
Jaydip Guha – Vice Chair
Huy Le – Treasurer
Tony Liu – Membership Chair
Jim Browning – University Liaison
Steve Groothuis, Gurtej Sandhu – Advisors

Manuscript Reviewers for WMED 2012

Amy Weatherly, Ben Millemon, Deepak Chandra Pandey, Dennis Montierth, Durga Panda, Eric Booth, Glen Hush, Howard Kirsch, Jake Baker, Jaydeb Goswami, Jian Li, Jim Hofmann, Kamal Karda, Kyle Kirby, Mark Fischer, Michael Andreas, Nishant Sinha, Puneet Sharma, Raghu Singanamalla, Rajesh Gupta, Rita Klein, Roy Meade, Sony Varghese, Sourabh Dhir, Steve Groothuis, Suresh Ramakrishnan, Tim Hollis, Vishal Saxena, and Wolfgang Mueller

IEEE WMED 2012 Technical Program

Friday, April 20, 2012 8:00AM-6:00PM

8:00AM	**Check In and Door Registration** *With Continental Breakfast*
8:30AM	**Welcome to WMED 2012** Room: Jordan D
9:00AM	**Keynote Address: "Bioelectronics"** *Prof. Rahul Sarpeshkar, Massachusetts Institute of Technology, Research Laboratory of Electronics* Room: Jordan D
10:00AM	**Break and Poster Setup**
10:15AM	**Invited Tutorials (Parallel sessions)** **Tutorial 1:** "Phase-Change Memory: Replacement or Transformational" *Dr. Chung Lam, Distinguished Engineer, IBM, T.J. Watson Research Center* Room: Jordan D **Tutorial 2:** "Practical Applications of Asynchronous Pipeline Circuits" *Brian Johnson, Distinguished Member of the Technical Staff, Micron Technology, Inc.* Room: Jordan AB
12:00 PM	**Buffet Luncheon** *Provided by WMED* Room: Hatch Ballroom
1:00 PM	**Invited Talk:** "Nano-Electro-Mechanical Memory Technology for Future Compact and Ultra-Low-Power Integrated Systems" *Prof. Tsu-Jae King Liu, IEEE Fellow, UC Berkeley* Room: Jordan D
1:45PM	**Invited Talk:** "Nanophotonic Interconnection Networks for Performance-Energy Optimized Computing" *Prof. Keren Bergman, IEEE Fellow, Columbia University* Room: Jordan D
2:30PM	**Invited Talk:** "Emerging Magnetic Memories" *Dr. William Gallagher, Senior Manager, IBM, T.J. Watson Research Center* Room: Jordan D
3:15PM	**Break**
3:30PM	**Contributed Papers (Parallel Sessions)** **Session 1**: Process and Devices Room: Jordan D **Session 2**: Solid State Circuits Room: Jordan AB
5:15PM	**Poster Session and Refreshments** Room: Hatch Ballroom

Contributed Papers

Session 1: Process and Devices
Room: Jordan D

3:30PM	"A Novel Method to Address ILD CMP Non-uniformity Issue for Advanced Memory Device Integration" *W. Wei, I. McDaniel and A. Jindal, Micron Technology Inc.; J. H. Ng, Micron Semiconductor*
3:50PM	"Numerical Simulation of Heat Generation during the Back Grinding Process of Silicon Wafers" *A. H. Abdelnaby, G. P. Potirniche, F. Barlow, and A. Elshabini, University of Idaho; S. Groothuis and R. Parker, Micron Technology, Inc.*
4:10PM	"Integration of Laser Control Technology with i-GLV to Directly Image High-Resolution Patterns for Advanced 3D Packaging" *H. Matsui, N. Kawamura, and J. Snow, Dainippon Screen Mfg. Co., Ltd.*
4:30PM	"Analytical Model of Deeply-Scaled Thyristors for Memory Applications" *D. Ventrice, P. Fantini, D. Betto, G. Carnevale and A. Benvenuti, Micron Technology Inc.*
4:50PM	"Step Deposition and Stabilizer Interaction in Electroless Nickel Bath for Bond Pad Metallization" *C. Tiwari, R. Nguyen, and Nick Phucas, Micron Technology, Inc.; B. B. Teo and T. Steneck, IM Flash Technology*

Session 2: Solid State Circuits
Room: Jordan AB

3:30PM	"On-Chip 3D Inductors using Thru-Wafer Vias" *G. VanAckern, R. J. Baker, A. J. Moll, and V. Saxena, Boise State University*
3:50PM	"An Algorithmic Study of DDR3 SDRAM On-Die Termination Switch Timings" *S.N. Wong, Micron Semiconductor*
4:10PM	"Multi-bit Continuous-Time Delta-Sigma Modulator for Audio Application" *R. Mohan, R. Koppula, S. Balagopal, and V. Saxena, Boise State University*
4:30PM	"Two Techniques to Reduce Gain and Offset Errors in CMOS Image Sensors using Delta-Sigma Modulation" *K. Yap and R. J. Baker, Boise State University*
4:50PM	"High Voltage Tolerant Stacked MOSFET in a Buck Converter Application" *S. Page, A. Wajda, and H. Hess, University of Idaho*

Poster Session and Refreshments
Room: Hatch

	"PMOS Device Performance Improvement by using Buried Contact Implants" *S. Qin, T. McDaniel, L. J. Liu, R. Burke, Y. J. Hu, and A. McTeer, Micron Technology, Inc.*
	"Deep Trench Patterning and Lift off Resist in Microfluidic Devices" *B. Pun, M. Mitkova, and P. Miranda, Boise State University; R. Zoller and M. Seibert, pSiFlow Technology, Inc.*
	"Nano-Ionic Conductive Bridge Memristors Based on Chalcogenide Glasses – Electrical Performance Characterization and Modeling" *M. R. Latif and M. Mitkova, Boise State University*
5:15PM	"Dependence of the Structure on Performance of Chalcogenide Glass Based Radiation Sensors" *M. S. Ailavajhala, P. Chen, M. Mitkova, and D. P. Butt, Boise State University; D. Olesky, Y. G. Velo, and H. Barnaby, Arizona State University*
	"Flexible Photovoltaics" *P. Salvador, M. Ostyn, and S. Parke, Northwest Nazarene University*
	"Measurement and Simulation of Hysteresis in LTCC Electron Hop Funnel IV Curves" *T. Rowe, M. Pearlman and J. Browning, Boise State University*

KEYNOTE ADDRESS

x

Keynote

Bioelectronics

Prof. Rahul Sarpeshkar, Massachusetts Institute of Technology

Abstract:

Nature is a great analog and digital circuit designer. She has innovated circuits in the biochemical, biomechanical, and bioelectronic domains that operate very robustly with highly imprecise parts, and with incredibly low levels of power. This talk will discuss how analog, RF, and bio-inspired circuits and architectures have led to, and are leading to, novel systems for ultra-low-power biomedical applications. Examples from systems for cochlear-implants for the deaf, brain–machine interfaces for the blind and paralyzed, and in bio-molecular circuits for systems biology and synthetic biology will be presented.

Speaker's biography

Rahul Sarpeshkar obtained Bachelor's degrees in Electrical Engineering and Physics at MIT. After completing his Ph.D. at CalTech, he joined Bell Labs as a member of the technical staff. He is currently on the faculty of MIT's Electrical Engineering and Computer Science Department, where he heads a research group on Analog Circuits and Biological Systems.

http://www.rle.mit.edu/acbs/

He holds over 25 patents and has authored more than 100 publications, including one that was featured on the cover of *Nature*. His book, *Ultra Low Power Bioelectronics: Fundamentals, Biomedical Applications, and Bio-inspired Systems* was released in February 2010 and contains a broad and deep treatment of the fields of bioelectronics and ultra-low-power electronics. He has won several awards for his interdisciplinary bioengineering research, including the Packard Fellow award given to outstanding faculty. He was a recent speaker at the 2011 'Frontiers of Engineering' conference hosted by the National Academy of Engineering.

Invited Tutorial

Phase Change Memory: Replacement or Transformational

Dr. Chung H. Lam, International Business Machines

Abstract

In this tutorial, a short account on the working principles of Phase Change Memory and its development will be introduced, followed by a comprehensive comparison with incumbent and other emerging memory technologies. Focus will be drawn to the technical requirements for the replacement of DRAM and NAND Flash with new memory technologies as these incumbent technologies are approaching their physical limits of conventional two-dimension scaling. Existing characteristics and the current state of the development of Phase Change Memory in the industry are examined to match the requirements of a replacement for DRAM and NAND Flash separately. The replacement scenario will be summarized with suggestions for further development. The transformational scenario begins with a study of the current landscape and future directions in enterprise computing and consumer electronics. We shall examine the opportunity of introducing a new memory system requiring holistic and collaborative efforts among the system and processor designers, as well as software engineers. Again, suggestions on further development directions for Phase Change Memory for the transformation scenario will be outlined.

Speaker's Biography

Chung Lam started his electrical engineering career at IBM Burlington as a memory circuit designer in 1978, upon graduating from the Polytechnic University of New York in 1978 with a B.Sc. In 1984 he was awarded the IBM Resident Study Fellowship and received his M.Sc. and Ph.D., both in Electrical Engineering, at Rensselaer Polytechnic Institute in 1987 and 1988, respectively. Since 1988, he had taken responsibilities in various disciplines of semiconductor research and development including circuit and device designs, as well as process integrations for memory and logic applications in IBM's Microelectronics Division. Currently, Chung is a manager at the IBM Research Division at the T.J. Watson Research Center leading the research on new non-volatile memory technologies. Chung was named an IBM Distinguished Engineer in 2007 and an IBM Master Inventor in 2009. He has about 160 issued US patents and has published more than 60 papers. He has been a member of the Technical Committee of the IEEE Non-Volatile Memory Workshop since 2001, and the IEEE International Memory Workshop since 2008 to 2011. Chung has also been a member of the Technical Committee of the VLSI-TSA since 2007, the ITRS Semiconductor Road Map Committee since 2008, and the IEDM 2011 Memory Technology Committee.

Invited Tutorial

Practical Applications of Asynchronous Pipeline Circuits

J. Brian Johnson, Micron Technology, Inc.

Abstract

Asynchronous circuit designs have shown great promise for overcoming synchronous timing overhead and power requirements of functionally equivalent, continuously-clocked circuit designs. In some instances, such as memory data path applications, speed independent, asynchronous pipelines are used to provide low-latency, high-throughput logic paths. Asynchronous pipelines use local handshaking to avoid the problems of clock skew, switching noise and switching power generated by global clock distributions found in synchronous circuit designs. Asynchronous pipelines that exhibit the property of *speed independence* provide functional tolerance to low supply voltage, and are modular in application, given proper adherence to interface protocol. The advantages of asynchronous circuits, relative to synchronous circuits, are not without cost. Asynchronous circuits can pose difficulties with testing, interfacing to synchronous environments, and a general lack of well-established, industry standard design and verification tools, which often forces asynchronous designs to be realized using full-custom design methodologies. This tutorial will present recent developments in the literature related to asynchronous pipeline protocols and circuits. Unique circuit constructs and applications will be presented with emphasis on interfacing asynchronous circuits to synchronous environments, controlling dynamic logic paths, and applications that call for the flexibility of varying forward path, inter-stage cycle times within an asynchronous logic pipeline.

Speaker's Biography

Brian Johnson earned a B.A. in English from Whitman College, Walla Walla, WA in 1988 and a B.S.E.E. from the University of Idaho, Moscow, ID in 1999. Following that, he completed his M.S.E.E. at the University of Idaho, Moscow, ID in 2004.

He has held various positions with Micron Technology, Inc. beginning in 1990, and is currently a Distinguished Member of the Technical Staff. He is a co-author of "DRAM Circuit Design: Fundamental and High-Speed Topics", IEEE Press, 2007, and has co-authored other papers published in IEEE journals. His current research interests include asynchronous circuit design and design of methods for timing and control circuits for computer memory sub-systems.

Mr. Johnson holds numerous patents related to DRAM design and CMOS circuit design.

Invited Talk

Nano-Electro-Mechanical Memory Technology for Future Compact and Ultra-Low-Power Integrated Systems

Prof. Tsu-Jae King Liu, University of California, Berkeley

Abstract:

As demand for mobile and compact computing devices increases in the digital information age, the need for low-power, low-cost nonvolatile memory (NVM) increases. To overcome the challenges of high programming voltages and/or currents, slow programming speed, and small sensing margin for conventional NVM technology, a simple electro-mechanical diode cell design recently has been proposed and demonstrated [1]. This presentation will review the cell structure and operation, and discuss the scalability and reliability of this technology. A nanoscale (sub-100nm) electro-mechanical (NEM) NVM technology is projected to offer significant advantages in speed (sub-ns programming time) and power consumption (< 1 fJ program/erase energy) over other established and emerging NVM technologies, and hence shows promise for future ultra-low-power memory applications.

Co-author: Wookhyun Kwon

[1] W. Kwon, J. Jeon, L. Hutin, and T.-J. K. Liu, *IEEE Electron Device Letters*, vol. 33, no. 1, 2012.

Speaker's Biography

Tsu-Jae King Liu is the Conexant Systems Distinguished Professor of Electrical Engineering and Computer Sciences and Associate Dean for Research in the College of Engineering at the University of California at Berkeley (UC-Berkeley). She received her B.S., M.S., and Ph.D. degrees in Electrical Engineering from Stanford University, and worked at the Xerox Palo Alto Research Center before joining the faculty at UC-Berkeley in 1996. Prof. Liu's awards include the DARPA Significant Technical Achievement Award in 2000 for development of the FinFET, the IEEE Kiyo Tomiyasu Award in 2010 for her contributions to nanoscale MOS transistors, memory devices, and MEMs devices, and the ECS Thomas D. Callinan Award in 2011 for excellence in dielectrics and insulation investigations. Her research activities are presently in nanoelectronic and nanomechanical devices for energy-efficient electronics. She has authored, or co-authored, over 400 publications, holds over 80 U.S. patents, and is a Fellow of the IEEE.

Invited Talk

Nanophotonic Interconnection Networks for Performance-Energy Optimized Computing

Prof. Keren Bergman, Columbia University

Abstract:

As the computational performance of microprocessors continues to grow, through the integration of an increasing number of processing cores, the interconnection network has become the central subsystem for providing the communications infrastructure among the on-chip cores, as well as to off-chip memory. This accelerated growth of multicore computational power is straining the capabilities of the interconnection network to deliver sufficient bandwidth to appropriately feed the processors, while severely widening the gap of the available off-chip memory access bandwidth. For many high-performance computing applications, the bandwidth available for both on- and off-chip communications can play a vital role in efficient execution, due to the use of data-parallel or data-centric algorithms. Electronic interconnected systems are increasingly bound by their communications infrastructure and the associated power dissipation of high-bandwidth data movement. Recent advances in chip-scale silicon photonic technologies have created the potential for developing optical interconnection networks that can offer highly energy efficient communications and significantly improve computing performance-per-Watt. This talk will examine the design and performance of photonic networks-on-chip architectures that support both on-chip communication and off-chip memory access in an energy efficient manner. Current challenges of inserting nanophotonic interconnect technologies into future computing systems will be discussed.

Speaker's Biography

Keren Bergman is the Charles Batchelor Professor and Chair of Electrical Engineering at Columbia University, where she also directs the Lightwave Research Laboratory.

http://lightwave.ee.columbia.edu/

She leads multiple research programs on optical interconnection networks for advanced computing systems, data centers, optical packet switched routers, and chip multiprocessor nanophotonic networks-on-chip. Dr. Bergman holds a Ph.D. from MIT and is a Fellow of the IEEE and of the OSA. She currently serves as the co-Editor-in-Chief of the IEEE/OSA Journal of Optical Communications and Networking.

Invited Talk

Emerging Magnetic Memories

Dr. William J. Gallagher, International Business Machines

Abstract

This talk will review emerging MRAM technologies, beginning with field-switched MRAMs that have been in product form for several years now, and extending to spin-torque MRAM under development now for higher density applications. Considerations for scaling to very small cell sizes will be discussed; in particular the importance of perpendicular magnetic anisotropy materials. Standard MRAM approaches involve the two-terminal magnetic tunnel junction device, that both stores the information and is used to read it out. A three-terminal magnetic device with potential advantages for speed will also be discussed, along with another approach dubbed "Racetrack" memory, which has the potential for three-dimensional shift-register like storage and very high densities.

Speaker's biography

William J. Gallagher joined IBM in 1979 after receiving his B.S. degree *summa cum laude* from Creighton University and his Ph.D. from MIT, both in physics. At IBM, he first worked on, and later managed, Josephson and Exploratory Cryogenics Research efforts. In 1989 Dr. Gallagher participated in the formation of the IBM-AT&T-MIT-founded Consortium for Superconducting Electronics and then served a Director of this consortium for six years. Since 1995, Dr. Gallagher has been leading MRAM development efforts at IBM. Currently, he is senior manager of Exploratory Magnetic Memory and Quantum Computing in IBM Research. Dr. Gallagher has over 160 technical publications, holds sixteen U.S. patents, and is a fellow of the IEEE and of the American Physical Society. He has served on the Executive Committee of the APS Forum on Physics and Society, on the Board of Directors of the Applied Superconductivity Corporation, and on study and review panels convened by universities and by organizations such as the National Research Council, the National Science Foundation, and the Office of Naval Research.

2012 IEEE WMED – Table of Contents

Welcome to WMED 2012 v
2012 IEEE WMED Management Committee... vi
WMED 2012 Technical Program vii
WMED 2012 Contributed Papers viii

Keynote Address
Bioelectronics... xi
> *Prof. Rahul Sarpeshkar, Massachusetts Institute of Technology*

Invited Tutorials
Phase Change Memory: Replacement or Transformational xii
> *Dr. Chung H. Lam, International Business Machines*

Practical Applications of Asynchronous Pipeline Circuits... xiii
> *J. Brian Johnson, Micron Technology, Inc.*

Invited Talks
Nano-Electro-Mechanical Memory Technology for Future Compact and Ultra-Low-Power Integrated Systems... xiv
> *Prof. Tsu-Jae King Liu, University of California, Berkeley*

Nanophotonic Interconnection Networks for Performance-Energy Optimized Computing xv
> *Prof. Keren Bergman, Columbia University*

Emerging Magnetic Memories.. xvi
> *Dr. William J. Gallagher, International Business Machines*

Technical Presentations
Process and Devices

- **Novel Method to Address ILD CMP Non-uniformity Issue for Advanced Memory Device Integration** 1
  W. Wei, I. McDaniel, A. Jindal, Micron Technology, Inc.; J. Hui Ng, Micron Semiconductor

- **Numerical Simulation of Heat Generation During the Back Grinding Process of Silicon Wafers**.. 5
  A. H. Abdelnaby, G. P. Potirniche, A. Elshabini, F. Barlow, College of Engineering, University of Idaho,; S.K. Groothuis, R. S. Parker, Micron Technology, Inc.

- **Integration of Laser Control Technology with i-GLV to Directly Image High-Resolution Patterns for Advanced 3D Packaging** 9
  H. Matsui, N. Kawamura, J. Snow, Dainippon Screen Mfg. Co., Ltd.

- **Analytical Model of Deeply-Scaled Thyristors for Memory Applications**........................... 12
  D. Ventrice, P. Fantini, D. Betto, G. Carnevale, A. Benvenuti, Micron Technology, Inc.

- **Step Deposition and Stabilizer Interaction in Electroless Nickel Bath for Bond Pad Metallization**... 16
  C. Tiwari, R. Nguyen, N. Phucas, Micron Technology, Inc.; B. B. Teo and Travis Steneck, IM Flash Technology

2012 IEEE WMED – Table of Contents

Solid State Circuits

- **On-Chip 3D Inductors Using Through Wafer Vias**... 20
  G. VanAckern, R.J. Baker, A.J. Moll, and V. Saxena, Boise State University

- **An Algorithmic Study of DDR3 SDRAM On-Die Termination Switch Timings**.................. 24
  S. N. Wong, Micron Semiconductor

- **Multi-bit Continuous-Time Delta-Sigma Modulator for Audio Application** 28
  R. Mohan, R. Koppula, S. Balagopal and V. Saxena, Boise State University

- **Two Techniques to Reduce Gain and Offset Errors in CMOS Image Sensors using Delta-Sigma Modulation** 33
  K. Yap, R.J. Baker, Boise State University

- **High Voltage Tolerant Stacked MOSFET in a Buck Converter Application** 37
  S. Page, A. Wajda, H. Hess, University of Idaho

Poster Sessions... 43

Advance Call for Papers... N/A

Author Index 47

A Novel Method to Address ILD CMP Non-uniformity Issue for Advanced Memory Device Integration

Wei Wei, Ian McDaniel, Anurag Jindal
Process R&D Department
Micron Technology
Boise, ID, USA

Jia Hui Ng
PEE300 Department
Micron Semiconductor Asia
Singapore, SG

Abstract—CMP non-uniformity has been an increasingly critical issue that needs to be addressed as memory device geometries continue to shrink. This issue is more prominent for ILD oxide CMP due to its poor controllability as compared to other CMP processes that can rely on stopping layers for effective endpoint detections and non-uniformity control. In this paper, we propose a novel approach to address the ILD oxide CMP non-uniformity issue by introducing a dual film polish concept. Experimental results show that this approach improves wafer non-uniformity and reduces scratches. Final device probe yield suggests that the approach is valid. A rate model is also proposed to elucidate the dual film CMP process.

Keywords: non-uniformity; ILD oxide CMP; dual film; rate model

I. INTRODUCTION

Chemical mechanical planarization (CMP) has been an enabling technology for IC manufacturing since the early 1990s. Inter-layer dielectric (ILD) oxide CMP was the first CMP step that was introduced to enable advanced lithography and dry etch processes at 0.8 μm technology node [1]. Over the years, ILD oxide CMP has been able to meet device integration requirements for multi-level metal interconnects at 0.35 μm, 0.13 μm, and 0.09 μm technology nodes for both logic and memory applications. However, due to the nature of the ILD oxide, there is no stopping layer and hence lacks an effective endpoint detection technology at CMP. It has been historically considered a "blind" CMP process, presenting tremendous challenges to achieving across wafer uniformity and defect control [2]. With semiconductor process technology advances to even smaller geometry, the within wafer non-uniformity issue of ILD oxide CMP, especially at the very edge of a wafer, becomes a top device yield killer for high volume IC manufacturing. Advanced process control technologies and hardware, such as *in-situ* polishing pressure control using multi-zone polishing heads and optical endpoint systems, have been developed for ILD oxide CMP process control for 0.09 μm technology and beyond [2]. However, current transistor number multiplication and geometry miniaturization to enable higher density and higher speed devices have outpaced semiconductor hardware advancements and breakthroughs.

A high polish rate CMP process can effectively reduce consumable usage, such as polishing pad and slurries, and hence be able to achieve high through-put and low cost of ownership (CoO) for high volume IC manufacturing. Boro-phospho-silicate glass (BPSG) removal rate can be tuned by doping with different concentrations of boron and phosphorus. The addition of phosphorus into oxide film will reduce the migration of alkali ions, and boron lowers glass transition temperature of the film, which allows it to reflow at a lower temperature and achieve higher polish rate and better local planarization performance [3]. A study by Bonner *et al.* showed that BPSG removal rate increases linearly with boron and phosphorus concentrations. It also reveals that removal rate is more sensitive to boron concentration when compared to phosphorus concentration [4]. Figure 1 shows the removal rate of 2% boron and 6% phosphorus doped blanket BPSG film compared with other oxide films. As shown in figure 1, the BPSG removal rate is more than three times higher than most other oxide films B-F [5].

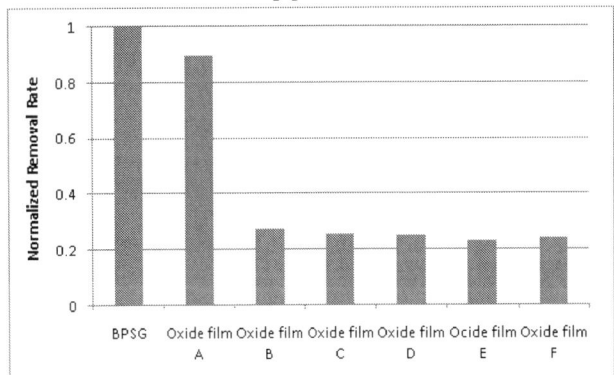

Figure 1: Oxide Removal rate comparison between BPSG film and other oxide films [5].

Albeit the fact that BPSG film is able to provide a tunable high removal rate, the increasing concern about BPSG ILD CMP is the ability to maintain a consistent across wafer non-uniformity, primarily at the wafer edge. This greatly reduces photolithography and dry etch process window at smaller device node and causes device yield loss. With current technology, the within wafer thickness variation of BPSG CMP process could be as high as 10% of total thickness removed. For example, the wafer thickness variation can be as high as 1,000 Å when removing 10,000 Å BPSG film after CMP. We observed this variation primarily at the wafer edge from 140 mm to 148 mm (with 2 mm edge exclusion). Using the advanced process control technologies such as *in-situ* polishing head zone pressure adjustment does not help in alleviating the edge non-uniformity issue due to the interactions between the edge zones [5].

The idea we propose here to address the BPSG ILD CMP non-uniformity issue is to utilize a lower removal rate (LRR) oxide film as a pseudo stopping layer for BPSG removal to achieve better within wafer non-uniformity. Since both films are oxide, this has minimum impact to the downstream etch processes.

II. SCHEMATICS AND THEORETICAL MODELS

A. Schematics

Figure 2 shows the concept of a dual layer oxide film scheme for BPSG ILD CMP.

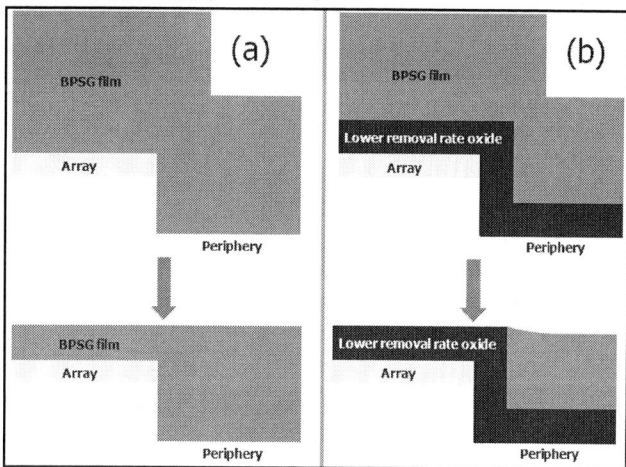

Figure 2: Schematics of (a) single BPSG and (b) dual film ILD CMP process.

The LRR oxide film will act as a pseudo stopping layer for BPSG CMP to mechanically slow down the polish process when hitting the film in array. The LRR oxide film that we have selected has about one third the removal rate when compared to BPSG film. The as deposited thickness of this oxide film is close to the desired remaining oxide thickness in array. We believe that the benefit of this dual film process will be manifested when we polish into a certain amount of LRR oxide film. As illustrated in Figure 2 (b), there will be a noticeable amount of dishing in BPSG oxide in periphery area due to faster BPSG removal when the LRR oxide in array is reached. Based on initial inline results, the dual film process has three times higher oxide dishing than single BPSG film process. Subsequent process optimizations are done to address this issue and achieve a comparable oxide dishing process.

B. Theoretical Model

A dual film ILD renders single-zone rate models useless for processes that incorporate Closed Loop Control (CLC), requiring development of a new rate model. Such a model can be explained in the following manner:

For a given oxide film, the blanket film polish rate is given by Preston's equation [6]:

$$R_b = KPV \qquad (1)$$

Where R_b is blanket film removal rate, P is applied pressure on a wafer at a contact point, V is the linear velocity of a pad relative to the wafer carrier, and K is a constant defined by the effects of different slurry chemistries and fluid dynamics, pad properties, pad conditioning, etc.

For a patterned wafer with topography, such as illustrated in Figure 2 (a) (higher array and lower periphery), array removal rate $R_a(t)$ is defined in equation (2) [7]:

$$R_a(t) = \frac{R_i - R_b}{1 + e^{(t-\mu)\times Eff}} + R_b \qquad (2)$$

Where R_i is experimentally determined initial rate in array, t is polish time, μ is topography factor, and *Eff* is planarization efficiency factor.

For periphery, removal rate $R_p(t)$, it is defined by equation (3).

$$R_p(t) = \frac{R_i - R_a(t)}{R_i - R_b} \times R_b \qquad (3)$$

Hence, topography removal rate $R_t(t)$ is the rate difference between array and periphery, as given in equation (4):

$$R_t(t) = R_a(t) - R_p(t) \qquad (4)$$

Initially, the removal rate of the array (R_a) is higher than the blanket rate (R_b) because the applied pressure is distributed across the array. Gradually, the array rate approaches the blanket rate R_b as the topography is planarized. On the other hand, the periphery removal rate R_p is initially lower than blanket rate R_b due to limited pad deformation, resulting in less contact with the pad. The periphery rate gradually converges with the array rate R_a as the topography is removed. This removal rate is simulated in Figure 3(a) for a single BPSG film scenario. Figure 3(b) shows the simulation result for a dual film scheme. As we can see, the array rate R_a and periphery rate R_p converge in the same way as a single zone BPSG film scheme initially but then the array rate R_a drops dramatically as the LRR oxide film is exposed in the array. Due to the significant removal rate difference between BPSG and LRR oxide (more than 3 to 1), the polish process is virtually slow down when the LRR oxide film is exposed in array and better across wafer non-uniformity is anticipated from lower removal rate process. Future study will be carried out to understand the rate model for dual film scheme when extensive polish is continued into the LRR oxide film.

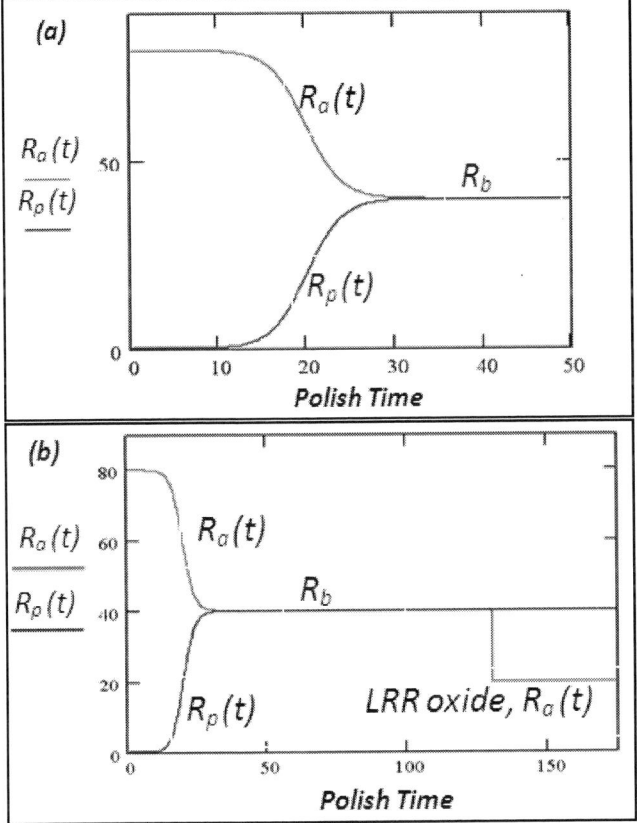

Figure 3: Polish rate convergence of array and periphery for (a) single BPSG film and (b) dual film ILD CMP.

III.   EXPERIMENTAL CONDITIONS

Four different experimental conditions were pursued: Group 1A, single BPSG film; Group 2B, BPSG film and LRR oxide film (200 Å less than nominal thickness); Group 3C, BPSG film and nominal LRR oxide film; Group 4D, BPSG film and LRR oxide film (200 Å more than nominal thickness). The nominal LRR oxide thickness was the desired oxide thickness remaining in array after CMP process. For the dual film groups, the total thickness of BPSG plus LRR oxide is identical to the single BPSG film stack in Group 1A. BPSG and LRR oxide film were deposited in the same CVD chamber sequentially. All 4 groups were polished with the same CMP recipe to hit the desired array oxide thickness as illustrated in Figure 2. ILD thickness was than measured across the wafer via scatterometry. Additionally, a scanning electron microscope (SEM) defect review was performed on each group to quantify and classify any defects. Yield results were also collected at probe.

IV.   RESULTS AND DISCUSSION

Figure 4 shows the normalized within wafer thickness profile comparison between the four experimental Groups.

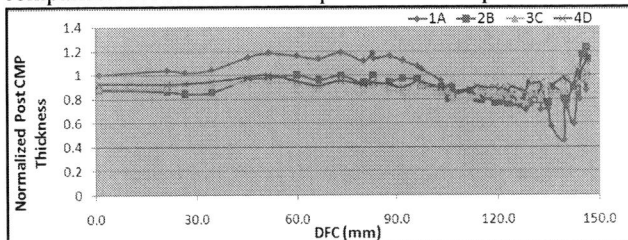

Figure 4: Within wafer thickness profile comparison between single BPSG group 1A and dual film groups 2B, 3C, and 4D.

As shown in Figure 4, the dual film groups 2B, 3C and 4D have a flatter profile (by ~ 20% to 50% range improvement) as compared to the single BPSG film group 1A, especially in two regions: 30 mm to 110 mm and 130 mm to 150 mm. Further uniformity improvements can likely be achieved past 145mm for the dual film groups through pressure optimization in the carrier zones.

Figure 5 shows normalized within-wafer-thickness range comparison between the four groups. Compared with group 1A, the dual film groups (2B, 3C, and 4D) show a 20%, 40% and 50% range improvement respectively. This further confirms that the LRR oxide film can effectively act as a pseudo stopping layer for BPSG film and slow the polish rate when the target thickness has been achieved. With such improvements in the CMP process, it is expected that downstream photo and dry etch processes will have less incoming variations and a wider process window. Detailed characterization of CD uniformity across wafer after photo and dry etch processes will be carried out in our next round of experiments. We believe the LRR oxide film uniformity is critical to this dual film process. The better the LRR oxide uniformity, the better the final uniformity will be for the process.

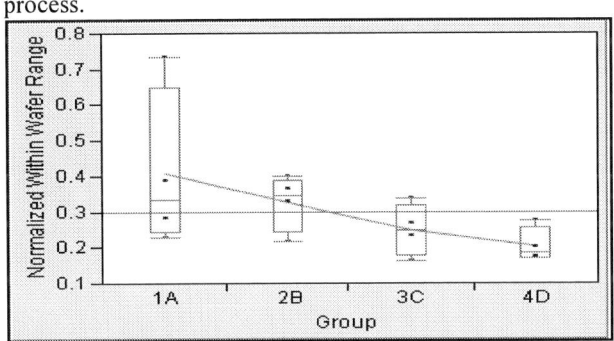

Figure 5: Normalized within wafer range comparison between single BPSG group 1A and dual film groups 2B, 3C, and 4D.

As discussed earlier, defects are becoming an increasingly challenging obstacle for ILD CMP as device geometries shrink below 0.05 μm. CMP arc scratches can cause shorts between adjacent contacts that will lead to device failure, either immediately after backend burn and stress test or gradually after Time Dependent Dielectric Break down (TDDB) [8]. With the dual film ILD CMP process, better scratch performance can be obtained than with traditional single film processes. Figure 6 shows groups 2B, 3C, and 4D generally achieved lower scratch counts than single BPSG film group 1A with group 2B and 3C showing the lowest scratch count.

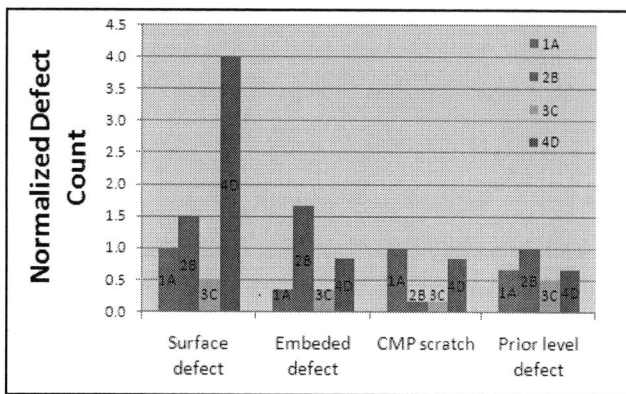

Figure 6: Defect comparison between single BPSG group 1A and dual film groups 2B, 3C, and 4D.

It is generally thought that ILD oxide CMP scratches are formed during the early stage of the process when a wafer first touches down on a pad [9]. Due to the high shear force created during the rotation of the wafer carrier and polishing pad, the agglomerated slurry particles tend to impinge into the oxide film and create large arc scratches. After the scratches are initiated, they propagate through the removal process. For the dual film process, our hypothesis is that when these scratches reach the interface between BPSG and LRR oxide film, they tend to slow down the propagation and are eventually polished away when the LRR oxide film is exposed. As the LRR oxide is exposed and the polish process continues, new scratches will form in LRR oxide film. Elevated scratch counts for group 4D, which polished down to 200 Å into the LRR oxide film, supports this hypothesis.

Figure 7 shows the normalized device functional yield per groups for this test. As we can see, groups 3C and 4D have considerably higher functional yield than group 1C. Group 2B shows comparable functional yield as group 1C but big variations from wafer to wafer within the group. Further tests need to be done to identify whether group 2B has any real yield toggle. Combined with CMP, scratches defect performance in Figure 6, we would choose group 3C to run more tests to confirm the yield benefit and dial in the dual film process.

V. CONCLUSION

A viable dual film approach has been demonstrated that improves post polish uniformity and decreases CMP scratches. Process optimization also showed dishing performance can be comparable to single film process. Future work will focus on process optimization and in depth defect characterization.

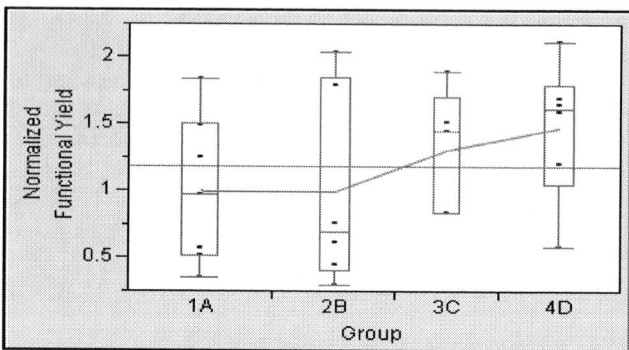

Figure 7: Functional yield comparison between single BPSG group 1A and dual film groups 2B, 3C, and 4D.

ACKNOWLEDGMENT

The authors would like to thank Sony Varghese for providing blanket oxide removal rate data and valuable suggestions during preparation of this paper and Shyam Ramalingam for providing fruitful feedback and support during the review process.

REFERENCES

[1] J. M. Steigerwald, "Chemical mechanical polish: the enabling technology", IEEE International Electron Devices Meeting (IEDM), pp. 1 – 4, 2008.

[2] T. Bibby, and K. Holland, "Endpoint Detection for CMP", Journal of Electronic Materials, 27 (10), pp. 1073-1081, 1998.

[3] S. J. Fang, S. Garza, and H. Guo, "Optimization of chemical mechanical polishing process for premetal dielectrics", Journal of the Electrochemical Society, 147(2), pp. 682, 2000.

[4] B. A. Bonner, B. Fishkin, J. David, C. Garretson and T. H. Osterheld, "Removal rate, uniformity and defectivity studies of chemical mechanical polishing of BPSG films", Material Research Society Symposium, 613, E8.6.1-E8.6.6, 2000.

[5] S. Varghese et al, "A Comprehensive Study of Nanomechanical Properties of Various SiO2-based Dielectric Films", IEEE WMED 2010.

[6] F. Tyan, "Non-uniformity of wafer and pad in CMP: kinematic aspects of view", Proceedings of American Control Conference, pp. 2047, 2005.

[7] I. McDaniel, "Construction and Utilization of a Double Asymptote First Order ODE for Process Modeling and Prediction of SIF CMP Processes", 2010 International Conference on Planarization/CMP Technology, P17, 2010.

[8] S. Jung, J. Uom, W. Cho et al, "A Study of Formation and Failure Mechanism of CMP Scratch Induced Defects on ILD in a W-damascene interconnect SRAM Cell" IRPS Proceedings, pp 42-47, 2001.

[9] S. Aytes, J. Armstrong, K. Mortensen et al, "Experimental investigation of the mechanism for CMP micro-scratch formation ", Proceedings of the 15th Biennial University, Government, Industry Microelectronics Symposium, pp. 107-109, 2003.

Numerical Simulation of Heat Generation during the Back Grinding Process of Silicon Wafers

A. H. Abdelnaby, G. P. Potirniche,
A. Elshabini, F. Barlow
College of Engineering
University of Idaho
Moscow, Idaho, USA
fbarlow@uidaho.edu

S.K. Groothuis, R. S. Parker
Process R&D Department
Micron Technology, Inc.
Boise, ID, USA

Abstract—The optimization of grinding parameters for silicon wafers is necessary in order to increase the reliability of electronic packages. Grinding is a mechanical process performed on silicon wafers during which heat is generated. The amount of heat generated affects the reliability of the wafer, and implicitly that of the final product. This paper describes the work performed to simulate the heat generated during a back grinding process for silicon wafers using the commercial finite element code ABAQUS. The grinding of a silicon wafer with a thickness of 60 μm mounted on a carrier wafer using bond adhesive material was simulated. The heat generated is caused by the friction between the grinding wheel and the backside of the silicon wafer. The computed temperature change due to friction in the wafer was compared with experimental and numerical values, and showed a good correlation. The numerical model developed can be used to better understand the local grinding temperature in the wafers and to estimate the effect of the grinding parameters on the temperature rise.

Keywords-Finite element; modeling; wafer; silicon; heat generation; grinding; friction

I. INTRODUCTION

The development of electronic devices is based on strict weight and size requirements. The thickness of silicon wafers affects the package size, so the thinner the wafer the smaller the package. The back grinding process is one of the common processes that are used to thin the silicon wafers. The grinding process is an energy intensive process, as it requires a large amount of energy per unit volume of material removed in comparison to metal cutting processes. During material removal, a portion of the mechanical energy is converted into heat, leading to a temperature rise in the contact grinding zone [1]. Our goal was to quantify the heat generated in the wafer in the form of a temperature rise during the grinding operation. The elevated localized temperature affects the mechanical properties of the materials involved in the process. Therefore it is important to build a model that has the capability of simulating the localized temperature, and has the ability to simulate the effect of the grinding parameters on its values.

Significant research efforts have been devoted to the development and improvement of the back grinding process for silicon wafers. Most of the studies have been experimental or analytical in nature, with few analyses considering numerical simulations that studied the wafer-wheel interaction

[2-5]. In order to better understand the details of the wafer grinding process, micron scale studies were previously used to clarify the internal stresses, strains, and deformations that take place in the wafer material during and after the grinding process[6-8].

In this paper a numerical study was performed to simulate and predict the heat generated from friction during the grinding process of silicon wafers as an extension to the previous simulations performed by the authors, which involved the study of the stresses induced in the back grinding process for bare silicon wafers as well as through silicon via wafers at the micron level [9].

II. FINITE ELEMENT MODEL

A. Model Geometry

The goal of this study is to predict the heat generated from friction during the back grinding process of silicon wafers in order to estimate the effect of the grinding parameters, such as grinding force, surface roughness and grinding speed on the grinding temperature. The numerical simulations involved varying operating parameters that influence this manufacturing process in order to determine the optimum operating conditions.

The commercial finite element software ABAQUS was used to simulate the heat generated during the grinding process. In the literature, there are different finite element grinding models, which can be categorized according to the scale of the modeling approach as macro-scale or micro-scale models. Macroscale models consider the overall wheel–workpiece interaction, which captures the aggregate effects of the abrasive wheel on the workpiece with no attempts to study the effect of the individual abrasive grain on the workpiece [8].On the other hand, microscale models focus on the individual grain–workpiece interactions, which can reveal the actual material removal mechanism. The macroscale modeling approach was used in this study in order to predict the wheel workpiece heat generation.

A two-dimensional model was built to simulate the entire wafer with a 300mm diameter. The wafer is fixed to the vacuum chuck along with the carrier wafer and backing tape. Out of the entire grinding wheel only three teeth were considered in the model. Each tooth had a 20mm width, they

978-1-4577-1735-2/12 $31.00 © 2012 IEEE

are in full contact with the wafer and moving along the wafer to simulate the grinding. A distributed force was applied on top of the grinding wheel to maintain contact between the teeth and the wafer, as shown in Fig. 1. The grinding force in the physical process was changed by changing the value of applied downward force on the wheel, as shown in Fig.2.

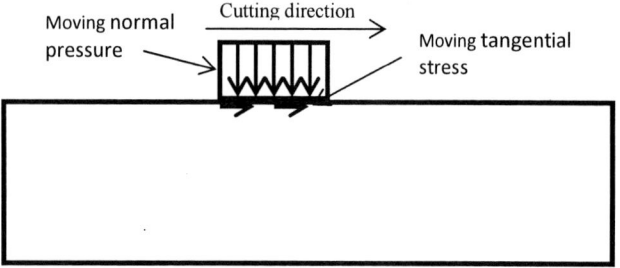

Figure1.Macro-scale grinding model

The simulations included three steps: (1) applying the grinding force, (2) moving the cutting teeth, and (3) applying the coolant. The cooling process was simulated as conduction heat transfer between the wafer surface, cutting teeth, and a coolant 2-D object placed in the finite element model above the wafer.

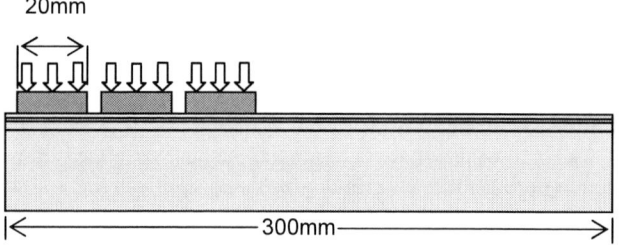

Figure 2. Finite element model used to simulate the macroscale grinding heat generation

The boundary conditions for the model were set as:
- The bottom of the model was pinned to simulate the attachment to the vacuum chuck.
- The initial temperature of all the parts in the model was set at 23°C to simulate the cooling effect before the grinding occurs.

The process parameters are:
- The coefficient of friction between the wafer and the grinding wheel was set to 0.5, while in the literature was reported to be between 0.45 and 0.6 according to the grinding stage [10].
- The grinding force was set to be 100N, it was reported to be between 67N and 200N [11].
- The coolant temperature was set at 23°C.
- The energy portion to be transformed to heat generated was chosen as 4%, as mentioned in [12].

The heat generated is due to the friction force between the grinding wheel and the silicon wafer. Therefore, the heat generated is function of the grinding parameters and this relation is governed by the following equation

$$Q = \mu PS \qquad (1)$$

where Q is the heat generated, μ is the coefficient of friction, P is the localized pressure, and S is the incremental slip.

The model consists of four materials: silicon as the wafer and carrier wafer material, bond adhesive as the backing tape materials, steel as the vacuum chuck material, and polymer as the grinding wheel material. Each of these materials was defined in the finite element simulation with a specific material model.

B. Material Models

Because of the crystalline nature of silicon <100>, orthotropic elasticity was chosen to model the elasticity of the material. Linear elasticity in an orthotropic material can be defined by nine independent elastic stiffness parameters. These parameters can be functions of temperature and other predefined fields, such as strain rate. In the present model there was no effect of the temperature on material behavior, because the cooling process maintained the wafer at 23°C. Also, the model was considered strain rate independent. In this case the stress-strain (σ - ε) relations take the form shown in equation 2 [13].

$$\begin{Bmatrix} \sigma_{11} \\ \sigma_{22} \\ \sigma_{33} \\ \sigma_{12} \\ \sigma_{13} \\ \sigma_{23} \end{Bmatrix} = \begin{pmatrix} D_{1111} & D_{1122} & D_{1133} & 0 & 0 & 0 \\ D_{1122} & D_{2222} & D_{2233} & 0 & 0 & 0 \\ D_{1133} & D_{2233} & D_{3333} & 0 & 0 & 0 \\ 0 & 0 & 0 & D_{1212} & 0 & 0 \\ 0 & 0 & 0 & 0 & D_{1313} & 0 \\ 0 & 0 & 0 & 0 & 0 & D_{2323} \end{pmatrix} \begin{Bmatrix} \varepsilon_{11} \\ \varepsilon_{22} \\ \varepsilon_{33} \\ \gamma_{12} \\ \gamma_{13} \\ \gamma_{23} \end{Bmatrix} \qquad (2)$$

A von-Mises plasticity model with an isotropic hardening criterion was used to simulate the plasticity of the silicon material. Silicon does not significantly deform in the plastic region before damage onset and fracture. The plasticity of the silicon was modeled using an effective stress-strain curve.

The grinding wheel material was modeled as a polymer material, and the fact that there are diamond grains embedded in the polymer was modeled by applying the corresponding coefficient of friction between diamond and silicon.

Bond adhesive is a hyper-elastic polymeric material used as a thin layer to model the tape on which the wafer is mounted to a carrier wafer, before the entire assembly is mounted on the vacuum chuck. In order to simulate the behavior of this material, the Moony-Rivlin model for hyperelastic materials was used. The model parameters were found from literature by curve fitting experimental data with the Mooney-Rivlin strain energy potential model used by ABAQUS [9].

III. RESULTS

The temperature in the model was investigated during the three steps mentioned earlier during the simulation.

During step one, while the grinding force was applied, there was no change in the model temperature. The reason for that is while applying the grinding force there is no parallel motion between the bodies, so there is no friction effect, therefore there no heat generation occurs.

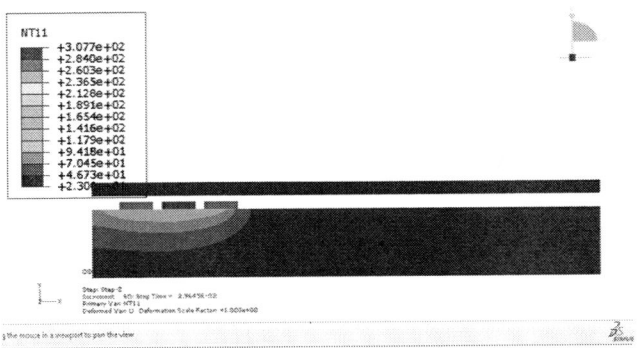

Figure 3.Temperature gradient in the beginning of step two

During step two, the grinding wheel was given a parallel motion, and the temperature of the model started to increase, the coolant has no effect instantaneously as it reaches the onset of boiling as reported in [15]. The change in temperature started in the contact between the two bodies, and then transferred to the entire assembly as shown in Fig. 3 and 4.

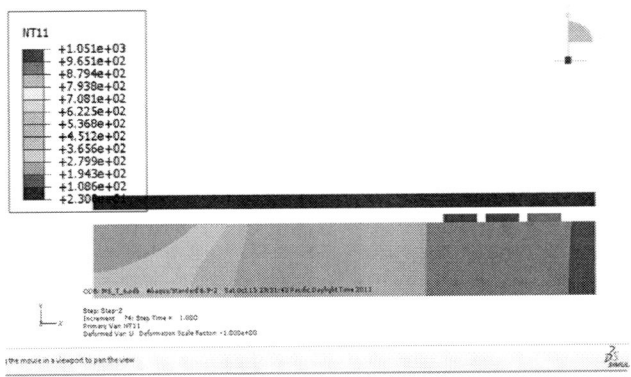

Figure 4.Temperature gradient by the end of step two

During step three, while the cooling was applied, the model temperature started to decrease to the ambient value of 23°C, as shown in Fig. 5.

From Fig. 5, it can be seen that the temperature decreases rapidly due to the effect of the coolant and also because it is localized. During the grinding process the temperature of the bulk of the wafer appears to be 23°C and also the coolant temperature doesn't change during the process. However, there is a local increase in temperature at the interface between grinding wheel and wafer, which affects the mechanical properties of the wafer while the grinding process occurs.

The contact temperature was measured with respect to the simulated grinding time during the grinding process, and it is shown in Fig.6.

A literature review was conducted to compare model predictions with previously published data. The present model results showed that the maximum temperature ranges between 900°C and 1050°C. These values agreed with the range of temperature found from experimental data in [14].

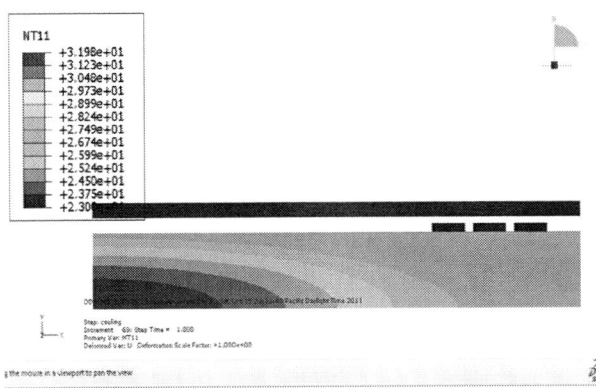

Figure 5.Temperature gradient at the end of step three

The grinding temperature values were also compared with those calculated by Jen and Lavine [15]. The model values agreed with the range of the maximum temperature that was calculated by them. It was also found that the wafer temperature expected to rise rapidly after a certain temperature due to the onset of nucleate boiling where the cooling effect of the coolant started to be inactive. Our model assumes that the coolant effect doesn't affect the grinding process, however the coolant could come into play after the grinding occurs to cool down the wafer to the room temperature. The onset of nucleate boiling differs with different coolants, however it ranges between 150°C and 320°C as found from the curves in the literature.

In Fig. 6 the horizontal line marks the value of the temperature of 950°C found in the literature compared to the simulation output, the figure also shows that there is a change in the behavior of the temperature rise after 300°C [15].

As a localized temperature of 1050°C during the grinding process is reached, the mechanical properties of the silicon material will be affected, which in turn will have an effect on the residual stresses induced in wafer after the grinding process. The localized temperature is also transferred through the wafer thickness to the front side. The wafer is then cooled down rapidly to the room temperature by the effect of the coolant.

The model was used to verify the effect of the grinding parameters on the heat generated. The coefficient of friction, grinding force, and the grinding velocity showed a remarkable effect on the output; however, the coolant temperature, the wafer thickness, and the elastic modulus of the grinding wheel did not show a remarkable effect on the temperature generated from friction during the process. The coefficient of friction is a surface property that cannot be controlled. Therefore reducing the grinding force or the grinding speed will reduce the heat generated during the process. The reduction of the heat generated will result in better a surface finishing and less thermal effects of the devices and metallization on the front side of the wafer.

IV. CONCLUSIONS

The development of electronic devices drives products towards slim, cheap, and high capacity packages. Development of such products increases the need of thinner wafers to decrease the thickness of the package. The back grinding of silicon wafers is one of the most common processes for thinning the wafers.

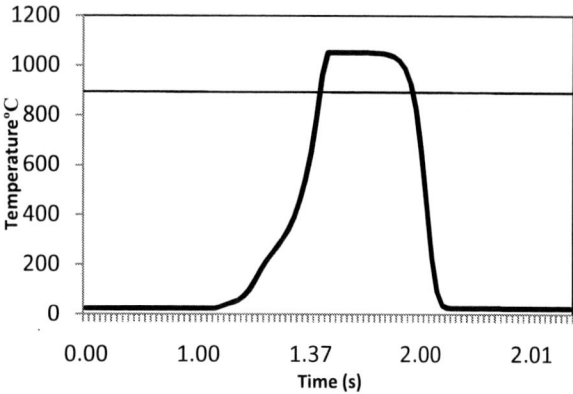

Figure 6.Change in temperature with time for a contact point

However the grinding process is an energy intensive process; it requires a large amount of energy per unit volume of material removed in comparison to metal cutting operations. The elevated localized temperature affects the mechanical properties of the materials involved in the process. Experimental studies are expensive and not always capable of accurately measuring the localized temperature in the wafer. Finite element simulations are an excellent tool to simulate the process and reduce the amount of resources needed to obtain the required data. In this study a model was built to simulate the heat generation during the grinding process of the wafers. The model developed was based on a macro-scale modeling approach. The model indicated that the temperature rise during the grinding process at the contact reaches a value of 1050°C. The temperature rise changes significantly the mechanical properties of the materials involved in the process. The model output agreed with the temperature range found in the literature. The temperature rise in the model output changed its behavior when it reached the value of 300°C, as the temperature started to rise rapidly because it reaches the onset of nucleate boiling and the cooling effect of the coolant is annihilated.

The present model is able to simulate the heat generated due to friction during the back grinding process of silicon wafers, and it can be used to develop a better understanding of the parameters affect the temperature values. The grinding force and the grinding velocity showed a remarkable effect on the output. Therefore reducing the grinding force or the grinding speed will reduce the heat generated during the process.

ACKNOWLEDGMENT

The authors would like to thank the Micron Foundation for the financial support of this work.

REFERENCES

[1] N. Alagumurthi, K. Palaniradja, and V. Soundararajan, "Heat generation and heat transfer in cylindrical grinding process -a numerical study," *The International Journal of Advanced Manufacturing Technology*, vol. 34, no. 5-6, pp. 474-482, May 2006.

[2] Z. Pei, G.R. Fisher, and J. Liu, "Grinding of silicon wafers: A review from historical perspectives," *International Journal of Machine Tools and Manufacture*, vol. 48, Oct. 2008, pp. 1297-1307.

[3] S. Chidambaram, Z.J. Pei, and S. Kassir, "Fine grinding of silicon wafers: a mathematical model for grinding marks," *International Journal of Machine Tools and Manufacture*, vol. 43, Dec. 2003, pp. 1595-1602.

[4] Y. Lin and S. Lo, "A study of a finite element model for the chemical mechanical polishing process," *The International Journal of Advanced Manufacturing Technology*, vol. 23, 2004, pp. 644-650.

[5] C.A. Chen and L. Hsu, "A process model of wafer thinning by diamond grinding," *Journal of Materials Processing Technology*, vol. 201, May. 2008, pp. 606-611.

[6] X. Chen and W. Brian Rowe, "Analysis and simulation of the grinding process. Part II: Mechanics of grinding," *International Journal of Machine Tools and Manufacture*, vol. 36, Aug. 1996, pp. 883-896.

[7] E. Ng and D.K. Aspinwall, "Modelling of hard part machining," *Journal of Materials Processing Technology*, vol. 127, Sep. 2002, pp. 222-229.

[8] D. Doman, A. Warkentin, and R. Bauer, "Finite element modeling approaches in grinding," *International Journal of Machine Tools and Manufacture*, vol. 49, Feb. 2009, pp. 109-116.

[9] A. H. Abdelnaby, G. P. Potirniche, F. Barlow, A. Elshabini, and R. Parker, "Finite element modeling of a back grinding process for Through Silicon Vias," in *2011 IEEE Workshop on Microelectronics and Electron Devices (WMED)*, 2011, pp. 1-4.

[10] "Material: Silicon (Si), bulk." [Online]. Available: http://www.memsnet.org/material/siliconsibulk/?keywords=silicon. [Accessed: 31-Oct-2011].

[11] X. H. Zhang, Z. J. Pei, and G. R. Fisher, "A grinding-based manufacturing method for silicon wafers: generation mechanisms of central dimples on ground wafers," *International Journal of Machine Tools and Manufacture*, vol. 46, no. 3-4, pp. 397-403, Mar. 2006.

[12] R. P. Upadhyaya and S. Malkin, "Thermal Aspects of Grinding With Electroplated CBN Wheels," *Journal of Manufacturing Science and Engineering*, vol. 126, no. 1, pp. 107-114, Feb. 2004.

[13] F. Ebrahimi and L. Kalwani, "Fracture anisotropy in silicon single crystal," *Materials Science and Engineering* A, vol. 268, Aug. 1999, pp. 116-126.

[14] T. Kuriyagawa, K. Syoji, and H. Ohshita, "Grinding temperature within contact arc between wheel and workpiece in high-efficiency grinding of ultrahard cutting tool materials," *Journal of Materials Processing Technology*, vol. 136, no. 1–3, pp. 39-47, May 2003.

[15] T.C. Jen, A.S. Lavine, A variable heat flux model of heat transfer ingrinding with boiling, *ASME J. Heat Transf.* 118 (1996) 463–470.

Integration of Laser Control Technology with i-GLV to Directly Image High-Resolution Patterns for Advanced 3D Packaging

H. Matsui, N. Kawamura and J. Snow

Dainippon Screen Mfg. Co., Ltd.
Kyoto, Japan 612-8486
hi.matsui@prp.screen.co.jp

Abstract— **A recently developed optical-write tool, incorporating a high-output laser with integrated grating light valve (i-GLV), has successfully been used to directly image high-resolution patterns useful for three-dimensional packaging schemes. This technology thus eliminates the need for masks for pattern creation. Automatic correction for any warpage from ultra-thin wafers is performed using a proprietary imaging technology. Different levels of dose intensity can be imaged on the resist in one exposure.**

Keywords-3D packaging, TSV, GLV, direct write

I. INTRODUCTION

Current device packaging schemes are striving for more functionality in smaller footprints via closely-packing or stacking the die in the z-direction using through-silicon vias, or TSV's. This 3D arrangement helps alleviate the interconnect bottleneck and enables the integration of more functionality, e.g. logic, memory, sensors, etc., in a more modular and scalable package as shown in Fig. 1 [1]. By connecting these stacked, thinned die with TSV's, the wiring lengths and power consumption are reduced and the speed of the device increased. It has recently been reported that Hybrid Memory Cubes using TSV's have a magnitude greater bandwidth compared to current high-tech devices at 10% of the footprint [2].

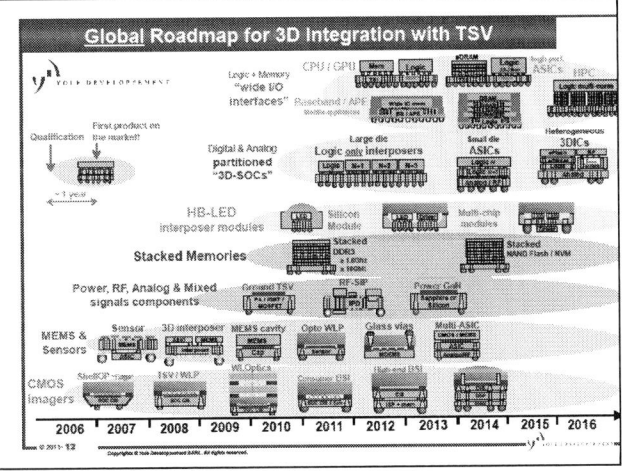

Figure 1.   3D Integration Technology Trend.

The thinning of the die utilized in these stacked devices has its own inherent processing challenges. Besides the difficulty in handling these ultrathin wafers, i.e. thicknesses 10-100 μm, the minuscule thickness allows warping that can easily result in lithographic distortion issues. While there exist some innovative lithographic processes using flexible exposure tools, a novel mask-less technique utilizing an optical direct-writing technique for 3D interconnect technology offers some inherent advantages.

II. EXPERIMENTAL

Experiments were performed on 300-mm wafers using a Dainippon Screen DW-3000 direct imaging exposure system [3]. This system incorporates two 16W YAG (355nm) lasers each modulated with an i-GLV, Integrated Grating Light Valve. Detection of alignment marks on the wafer top-side is accomplished by visible light and on the back-side by a reflective IR alignment system. Resolution is ≤ 3 μm (lines and spacers) and top-side and back-side alignment overlay is ≤ 1 μm (3σ). A detailed description of the tool and data processing flow has been previously reported [4]. Resists were from Tokyo Ohka Kogyo Co., Ltd. Cross-sectional SEM's were performed using a Hitachi S-4800.

III. RESULTS AND DISCUSSION

A. Local Alignment Function

A 300-mm wafer was intentionally distorted to simulate a warped substrate that might be encountered during 3D lithography. Four global and five local alignment positions were located across the wafer surface. The dislocation of each chip was calculated based on the local alignment measurement, and then the result was translated to the shift component in X/Y. The global wafer scaling was also adjusted and added with the local alignment. This ensured that the dislocation of every chip was well adjusted. Using global alignment only, errors in the mismatch of the exposure positions over 504 measurement points/wafer were -2.11 μm (3σ of 6.95 μm) in the x-direction and 0.90 μm (3σ of 5.68 μm) in the y-direction. The 3σ values were decreased by ~3.5x by incorporating the local alignment with the global alignment as shown in Fig. 2 and Table I.

B. TSV Patterning

A 300-mm wafer coated with 3.4 μm of TCIR ZR8800 resist was imaged at 130 mJ/cm². After development, the top of the hole in the resist measured 3.62 μm and the diameter at the bottom was 3.42 µm. Subsequent deep reactive-ion etching using the Bosch process provided the 60-µm deep TSV hole as shown in Fig. 3.

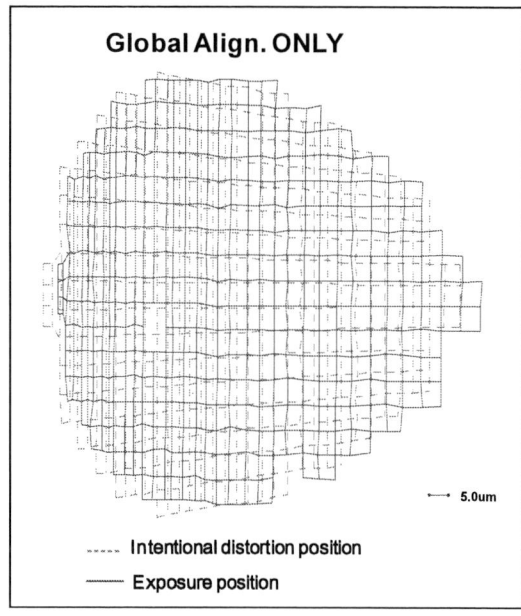

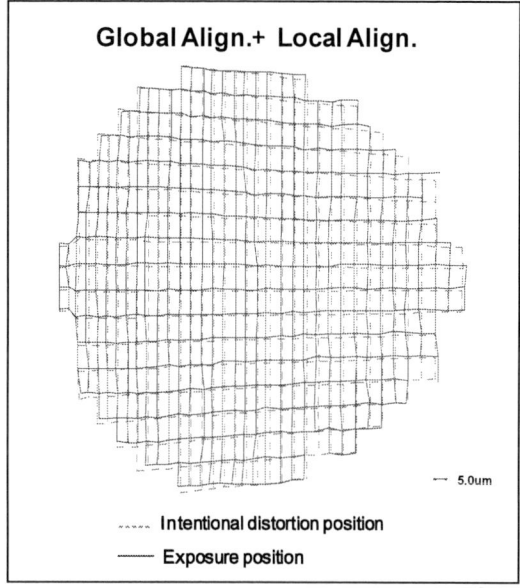

Figure 2. Alignment Accuracy Results from Distorted Wafer.

TABLE I. ACCURACY IMPROVEMENT BY LOCAL ALIGNMENT CORRECTION

	Global Alignment Only		Global + Local Alignment	
	x	y	x	y
Average	-2.11 µm	0.90 µm	-0.59 µm	0.84 µm
3σ	6.95 µm	5.68 µm	1.93 µm	1.64 µm

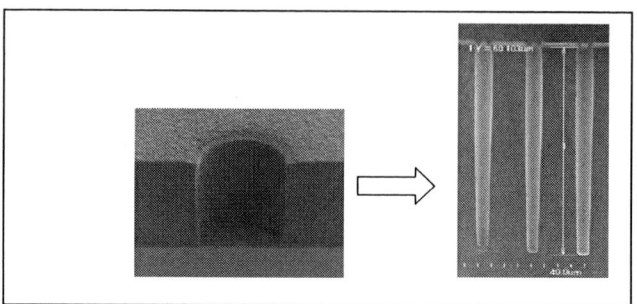

Figure 3. TSV Patterning of Wafer.

C. Gray-Scale Exposure

Unlike existing micro mirror devices such as DMD, a GLV is able to reflect several different levels of light intensity, i.e. not just on and off. By using this unique function, it is possible to expose different dose levels onto the substrate with a single exposure operation. A 300-mm wafer was coated with 20-µm of negative photoresist and exposed at 100 mJ/cm^2 and 200 mJ/cm^2 using the gray-scale function to intentionally generate sloped vias as shown in Fig. 4. This function was also used to create vias and RDL lines at the same time as shown in Fig. 5.

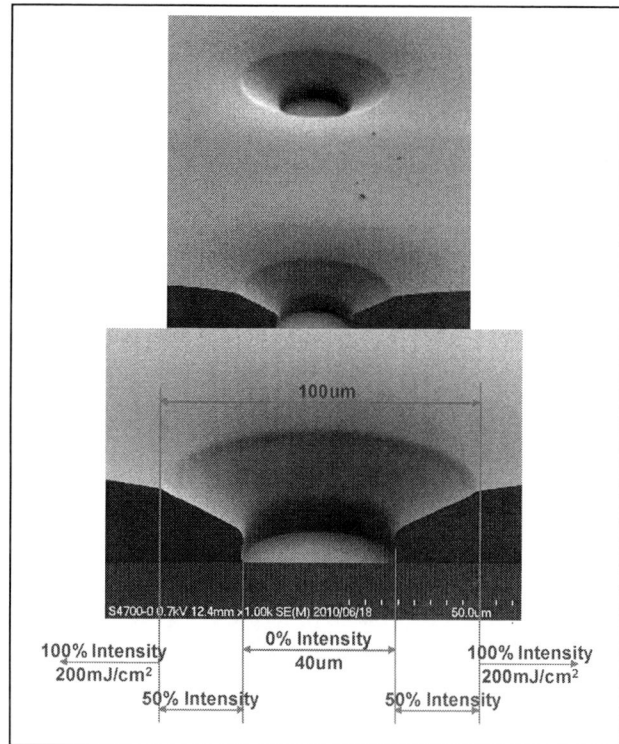

Figure 4. Creation of Sloped Vias using a Single Exposure.

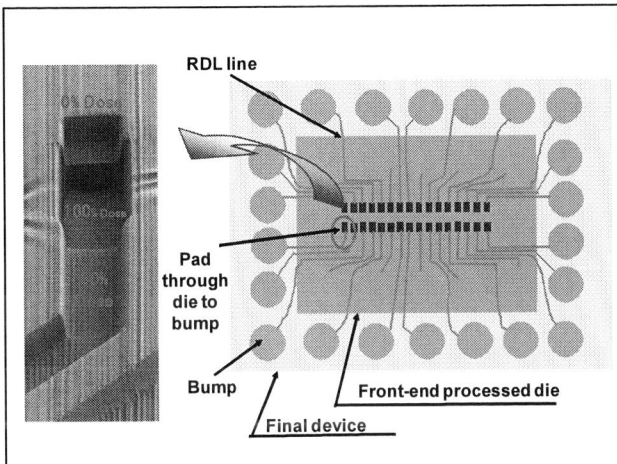

Figure 5. Single-Exposure Creation of Vias and RDL Patterns.

SUMMARY

The successful implementation of a DW-3000 optical write tool for creating TSV's without the use of an exposure mask has been demonstrated. The ability to image without a mask can enable rapid evaluation of new patterns and processes without the associated cost and delay of acquiring and maintaining new masks. Laser writing has the potential to scribe each chip with its own identification number, which will provide ultimate traceability. Gray-scale exposure provides the opportunity to create sloped vias and multiple patterns in a single exposure. When wafers are thinned to tens of microns, wafers tend to deform with non-linear shape from the thinning process or from the glass/plastic substrate to which they are bonded. The DW-3000 tool can accommodate wafer bowing up to 500 µm, and correct for this wafer distortion using the global plus local alignment functions. Improvements in the resolution and overlay accuracy specifications are continuously in process to meet the challenging 3D roadmap requirements. This tool will prove useful to handle the complex 3D multilayer substrates.

REFERENCES

[1] J. Perkins, "3DIC & TSV Interconnects – Market Update & Infrastructure Readiness", RTI Conference, December 2011. Figure provided and used with permission from Yole Devellopement.

[2] "IBM to Produce Micron's Hybrid Memory Cube in Debut of First Commercial, 3D Chip-Making Capability", (2011); http://www-03.ibm.com/press/us/en/pressrelease/36125.wss.

[3] "Dainippon Screen Enters Market for Semiconductor Post-processing Exposure Systems", Doc. No. NR111125E, (2011); http://www.screen.co.jp/eng/press/pdf/NR111125E.pdf.

[4] T. Azuma, M. Sekiguchi, M. Matsuo, A. Kawasaki, K. Hagiwara, H. Matsui, N. Kawamura, K. Kishimoto, A. Nakamura and Y. Washio, "A Novel Lithography Process for 3D (Three-Dimensional) Interconnect Using an Optical Direct-Writing Exposure System", Proc. SPIE 7637, 763710 (2010); doi:10.1117/12.846013.

Analytical model of deeply-scaled thyristors for memory applications

D. Ventrice, P. Fantini, D. Betto, G. Carnevale, A. Benvenuti

R&D Technology Development, Micron Inc., Agrate Brianza, Italy

dventric@micron.com

Abstract— In this paper we propose an analytical model for the static operation of thyristor, aiming at clarifying the basic physics involved in the switching from the low-conductance to the high-conductance state of the device. Modeling results are compared to TCAD numerical simulations, showing that the analytical calculations can nicely reproduce the main features of the current-voltage device characteristics.

Index Terms: thyristor, compact model, TCAD, T-RAM.

I. INTRODUCTION

The possibility to develop a high-performance volatile memory device making use of a deeply-scaled thyristor as the storage element has been proposed by Nemati et al. [1] with the thin-capacitive-coupled thyristor (TCCT) cell. In order to understand the achievable performance of this technology, an in-depth understanding of the thyristor characteristics at the nanoscale is mandatory, requiring, first of all, a clear picture of the basic physics involved in device switching.

In this work, we present a simple analytical physical based model for the static operation of non-gated thyristor, that well describes its electrical characteristics. We investigated the conditions for device switching from a low conductance to a high conductance state, considering the electron and hole fluxes through the p-n-p-n junctions of the device. Then we compared our modeling results are with those coming from numerical TCAD simulations, showing their nice agreement and confirming the validity of the proposed model for a first-order evaluation of TRAM characteristics.

II. ANALYTICAL MODEL

We studied the structure reported in Fig. 1 considering a uniform doping in each region. For our analytical model, we considered the minority diffusion currents through the quasi-neutral regions of the device, using the well-known diffusion equations. Boundary conditions for minority concentrations at the limit of quasi-neutral regions are given by p-n junction theory as functions of the voltage drops V1, V2 and V3 on the J1, J2 and J3 junctions, respectively. Because of the very small widths of the quasi-neutral regions in comparison with the carrier diffusion lengths, we neglected generation/recombination effects in this regions, resulting in minority carrier concentrations varying linearly in space. With the coordinate system of Fig.1, for example, the hole concentration in region n1 was calculated as:

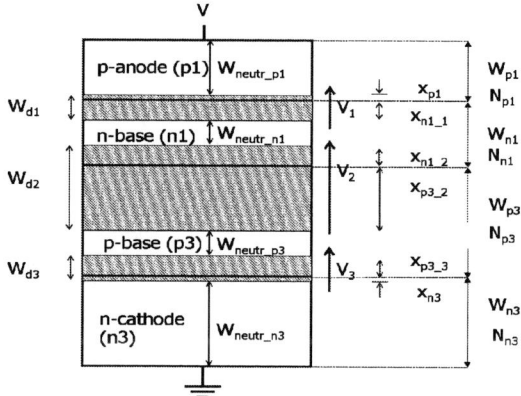

Fig. 1. Schematic structure of the considered thyristor, with the parameters related to the three junctions.

$$p(0) = p_{n_0} \exp(V_1/V_T)$$

and

$$p(W_{neutr\_n_1}) = p_{n_0} \exp(-V_2/V_T)$$

where $V_T = kT/q$ is the thermal voltage and $p_{n_0} = n_i^2/N_{n_1}$, being n_i the intrinsic concentration of carriers and N_{n1} the doping concentration in the n1 region. In p1 and n3 regions, we assumed equilibrium concentration n_{p0} and p_{n0} at the anode and cathode contacts. If we substitute boundary conditions into diffusion equations, we obtain:

$$
\begin{aligned}
J_{n\_1} &= \frac{qD_n n_i^2}{W_{neutr\_p_1} N_{P_1}} \left[\exp\left(\frac{V_1}{V_T}\right) - 1 \right] \\[2mm]
J_{p\_1} &= \frac{qD_p n_i^2}{W_{neutr\_n_1} N_{n_1}} \left[\exp\left(\frac{V_1}{V_T}\right) - \exp\left(\frac{-V_2}{V_T}\right) \right] \\[2mm]
J_{n\_3} &= \frac{qD_n n_i^2}{W_{neutr\_p_3} N_{P_3}} \left[\exp\left(\frac{V_3}{V_T}\right) - \exp\left(\frac{-V_2}{V_T}\right) \right] \\[2mm]
J_{p\_3} &= \frac{qD_p n_i^2}{W_{neutr\_n_3} N_{n_3}} \left[\exp\left(\frac{V_3}{V_T}\right) - 1 \right]
\end{aligned}
\tag{1}
$$

978-1-4577-1735-2/12 $31.00 © 2012 IEEE

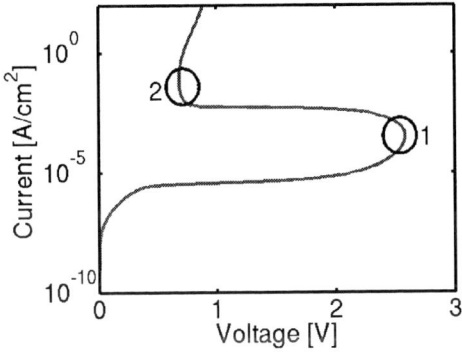

Fig. 2. I-V curve obtained with the model presented in Sec. II in the forced current mode

Index p/n_j refers to holes and electron currents exiting J1 or J3 junctions with j=1, 3 respectively, W_{neutr_p/n_j} are the widths of quasi-neutral regions, and $D_{n/p}$ are the diffusion coefficients for electrons and holes.

In addition to currents through quasi-neutral regions, we considered trap-assisted generation/recombination processes in the depletion layers of device junctions according to the Shockley-Read-Hall (SRH) formalism [2], [3]:

$$J_{sc} = (qn_i W_d / 2\tau)[\exp(V/2V_T) - 1]$$

where V is the forward voltage drop on the considered depletion region and τ is the carrier lifetime [4], assumed as a fitting parameter.

Both SRH generation/recombination and impact ionization were taken into account in reverse biased junction. Therefore, if an electron current density J_n enters the depletion region of a reversely biased junction, it is multiplied by a factor M, then providing a holes current density $(M-1)J_n$ at the other side of the junction. For the generation current in the depletion region, we assumed a localized generation of electron-hole pairs in the center of the junction: an electron flows towards the anode contact, generating M/2 hole-electron pairs, while an hole flows towards cathode contact, generating M/2 pairs too. As a result we have both electrons and holes currents multiplied by a factor M, which can be calculated according to the empirical formula:

$$M = \left[1 - \left(\frac{-V}{V_{BD}} \right)^n \right]^{-1} \qquad (2)$$

where $n \cong 5$ and V_{BD} is the breakdown voltage of the p-n junction. Both n and V_{BD} are used as parameters for the model.

In the device structure of Fig.1, when anode is positively biased and cathode is grounded, J1 and J3 junctions are initially forward biased, while J2 is reverse biased. Neglecting generation/recombination in quasi-neutral regions, conservation of holes and electrons currents is implied. This is equivalent to assume that the base current of the upper pnp

bipolar transistor is supplied by J3 and generation electrons current, since there is no base contact (open-base BJT) [5]. In summary, the basic equations of the model are:

$$J_{n\_1} + J_{sc\_1} = M\left(-J_{sc\_2} + J_{n\_3}\right) + (M-1)J_{p\_1} \qquad (3)$$

for the continuity of electrons current and:

$$J_{p\_3} + J_{sc\_3} = M\left(-J_{sc\_2} + J_{p\_1}\right) + (M-1)J_{n\_3} \qquad (4)$$

for the continuity of holes current.

Then, the total current Jtot is:

$$J_{tot} = M * J_{p\_1} + M * J_{n\_3} - M * J_{sc\_2} \qquad (5)$$

while the total voltage drop between anode and cathode is simply the sum of the three voltage drops on the junctions:

$$V_{tot} = V_1 + V_2 + V_3 \qquad (6)$$

Since we have 3 unknown variables, V_1, V_2 and V_3, we need three equations to solve the system, the first two being the continuity equations (3) and (4). The third equation can either be Eq. (5) or (6) depending on the external forcing of anode voltage or total current flowing in the cell.

III. SIMULATION RESULTS: EXPLANATION OF THE SWITCHING

The non-linear equations system giving the currents and voltages across the p-n-p-n structure of Fig.1 was solved using MATLAB. For each iteration of the solver, new values of the depletion layer widths and quasi-neutral regions widths were calculated. In this way, we took into consideration effects such as base punch-through and Early effect for the two BJT transistors. Fig. 2 shows the results of the simulation obtained with the model introduced in the previous section: in order to reproduce the I-V characteristic with the typical negative differential resistance (NDR) feature the forced current mode was employed. The two points highlighted in the figure are the initial and the end point for the NDR region. Fig. 3 shows the evolution of the voltage drop on each junction when changing the current forced through the device.

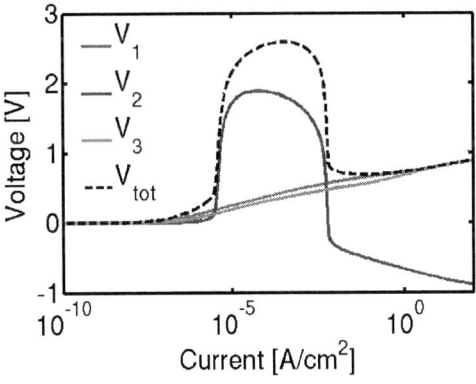

Fig. 3. Voltage drop on each junction as a function of the total current in the thyristor

At low current levels the total voltage is almost equal to the reverse voltage drop on J2, while after the switching all the junctions are forward biased and all the voltage drops are almost equal in absolute value.

From the analysis of these results, the switching of the I-V curve is obtained when:

$$\frac{\partial(V_1 + V_3)}{\partial J} = -\frac{\partial V_2}{\partial J} \qquad (7)$$

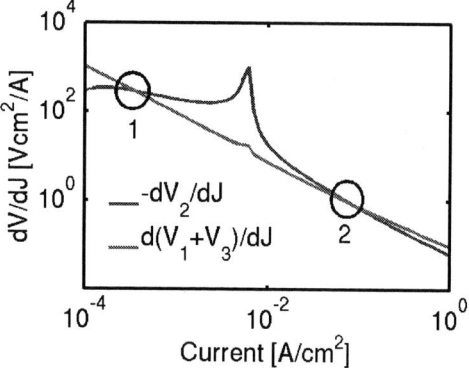

Fig. 4. Switching condition for the thyristor: the two points highlighted are the ones shown also in Fig. 2, corresponding to the switching of the device.

i.e., when the increasing rate of V_1 and V_3 is equal to the decreasing rate of V_2. This is shown in Fig. 4, where the two terms in (7) are represented: the two points at which the curves cross, are the two points highlighted in Fig. 2, at the beginning and at the end of the NDR region.

IV. SIMULATION RESULTS: COMPARISON WITH TCAD

Fig. 5 compares the simulation results obtained by means of the analytical model of the previous sections and of TCAD tools for the structure sketched in Fig. 1. The parameters of the analytical model were chosen in order to fit TCAD simulation. Fig. 5(b) also reports the comparison between TCAD and MATLAB simulations for the band structure of the cell at different points of the I-V characteristics; in the analytical model, the band structure is calculated assuming that the entire voltage drop of each junction is the voltage drop on the depletion layer. Point 1, shown in the left part of Fig. 5(b), is taken before the forward breakover and displays that J1 and J3 are forward biased while J2 is reverse biased. In this condition the applied voltage V_{tot} is almost equal to V_2, as a quite small voltage drop on J1 and J3 is enough to sustain the reverse current of J2. Instead point 2, shown in the right part of Fig. 5(b), is taken at the end of the snap-back region of the I-V curve, and displays that, in the on state, also J2 is forward biased, in agreement with the thyristor theory: the obtained current is many orders of magnitude higher than the one obtained in the lower part of the curve.

Other two versions of the analytical model have been developed. The first one considers different multiplication factors M_p and M_n for holes and electrons respectively,

calculating them with the van Ovestraeten – de Man model [6]: the actual multiplication factors are calculated by integrating the ionization coefficients in the depleted region, taking into consideration also their dependence on the electric field. The second version uses the approximate model proposed by Hurkx [7], [8], [9] to include the band-to-band tunneling and trap-assisted tunneling in the reverse biased junctions. In this case, these contributions are modeled as additional

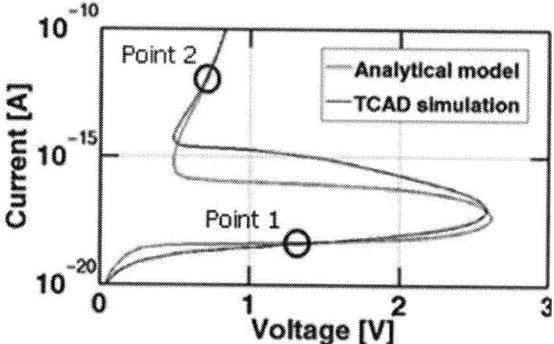

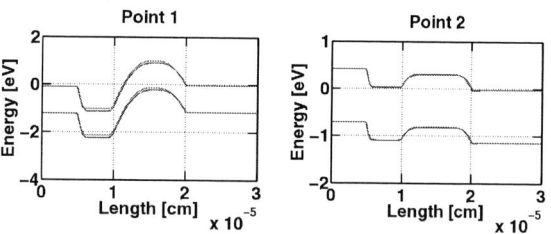

Fig. 5. Comparison between the proposed analytical model and TCAD simulation. (a): I-V curve, (b): energy band structures at point 1 (left) and point 2 (right) of the I-V curve.

generation/recombination rates in the depletion layer. This model also allows to remove the approximation that the generation current is localized at the center of the junction: the multiplication factor for the generated electron/hole pairs is computed for each point in the depleted region and then integrated. Anyway, the basic equations and the model framework remain the same (equations (3) to (6)), and just the calculation of the parameters changes.

A fundamental step in order to allow the analytical model to describe the TCCT behavior is to include the effect due to the gate coupling. For this purpose we performed several TCAD simulations of the static Ianode(Vanode) characteristic both with a fixed doping profile changing the gate voltage and without the gate changing the doping concentrations.

Due to the MOS-like control of the p-base the gate voltage changes the I(V) characteristic of the device from the typical S-shaped curve with a forward breakover voltage of a thyristor (when the p-base is in accumulation) to a diode-like behavior (when the p-base is in inversion). Qualitatively the same dependence can be obtained describing the p-base

978-1-4577-1735-2/12 $31.00 © 2012 IEEE

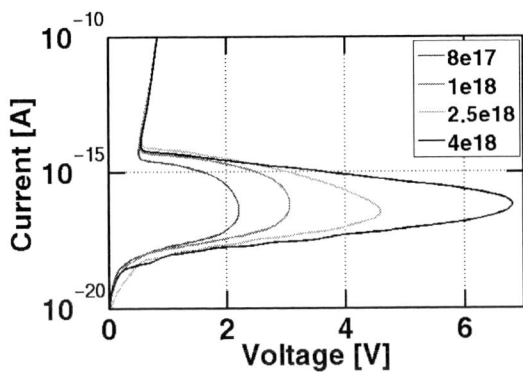

Fig. 6. TCAD simulations of the gate coupling effect via p-base effective doping.

accumulation or inversion by means of an effective uniform doping. This is illustrated in Fig.6, which shows through a TCAD simulation the modulation of the forward breakover voltage as a function of such effective doping. Please note that in this case tunneling generation was not included to emphasize the effect of p-base punchthrough on the forward breakover voltage. Following the same idea the forward breakover voltage can be modulated also in the proposed analytical model of thyristor by varying the p-base effective doping. Furthermore, we plan to introduce the gate effect to mimic the real cell behavior by means of a physics-based analytical representation.

CONCLUSIONS

In conclusion, we developed a new analytical model for the vertical non-gated thyristor device able to describe its static electrical behavior. We plan to improve the model by adding the gate effect and AC components. In this sense, it represents the initial building block for a complete SPICE-like model useful to design next generation capacitor-less DRAM integrated circuits.

ACKNOWLEDGMENT

The authors would like to thank C. Monzio, A. Spinelli and A. Maconi from Politecnico di Milano for the fruitful discussions.

REFERENCES

[1] "A Novel Thyristor-based SRAM Cell (T-RAM) for High-speed, Low-Voltage, Giga-scale Memories", Farid Nemati and James D. Plummer, IEDM, 283-286, 1999.

[2] C. Sah, R. N. Noyce, W. Shockley, "Carrier Generation and Recombination in p-n Junctions and p-n Junction Characteristics", Proceedings of the IRE, Vol. 45, N. 9, September 1957.

[3] P. U. Calzolari and S. Graffi , "A Theoretical Investigation on the Generation Current in Silicon p-n Junctions Under Reverse Bias", Solid State Electronics , Vol. 15, 1972.

[4] C. T. Sah, R. N. Noyce and W. Shockley, "Carrier generation and recombination in p-n junction and p-n junction characteristics", *Proc IRE*, vol. 45, pag. 1228, 1957.

[5] J. F. Gibbons, "A critique of the theory of p-n-p-n devices", *IEEE Trans. on Electron Devices*, vol. 11 issue 9, pag. 406-413, 1964.

[6] R. van Overstraeten and H. de Man, "Measurements of the ionization rates in diffused silicon p-n junctions", *Solid State Electronics*, vol. 34 issue 1, pag. 230-238, 1957.

[7] G. A. M. Hurkx, H. C. de Graaff, W. J. Kloosterman and M. P. G. Knuvers, "A new analytical diode model including tunneling and avalanche breakdown", *IEEE Trans. on Electron Devices*, vol. 39 issue 9, pag. 2090-2099, 1992.

[8] G. A. M. Hurkx, D. B. M. Klaassen and M. P. G. Knuvers, "A New Recombination Model for Device Simulation Including Tunneling", IEEE Transactions on Electronic Devices, Vol. 39, N. 2, February 1992.

[9] G. A. M. Hurkx, D. B. M. Klaassen, F. G. O'Hara and M. P. G. Knuvers, "A New Recombination Model Describing Heavy-Doping E ects and Low-Temperature Behaviour", Electron Devices Meeting, IEDM Technical Digest, IEEE International, 1989.

Step Deposition and Stabilizer Interaction in Electroless Nickel Bath for Bond Pad Metallization

Chandra Tiwari, Robert Nguyen,
and Nick Phucas
Micron Technology Manassas, VA,
USA
ctiwari@micron.com

Bee BEE Teo
IM Flash Technology
Singapore

Travis Steneck
IM Flash Technology
Lehi, UT, USA

Abstract—Stabilizers are key components of electroless nickel baths. Therefore, understanding the limitations and characteristics of stabilizers is very important for consistently achieving desired plating results. Higher concentrations of stabilizers cause defects like skip and step plating, whereas lower concentrations can lead to short bath life and bath decomposition. This paper studies the step deposition and stabilizer interaction on electroless nickel bath.

Keywords- Electroless nickel; stabilizer; skip plating; step plating ; anisotropic deposition; bond pad; defects

I. INTRODUCTION

Since the invention of integrated circuits, the semiconductor industry has been on relentless path of scaling, forcing innovation and change. This has led to the introduction of novel technologies, processes, and materials. Electroless processes, which are ubiquitously used for coating metals and alloys in many industries, have found applications in chip manufacturing in recent years. CoWP, NiP, Pd, Au, and Cu are some of the metals/alloys used as barriers, interconnects, or as a bonding surface [1,2,3]. Electroless nickel is used for under bump metallization (UBM) in flip chip technology [4]. It also has found application in bond pad metallization in copper interconnects.

In the last decade, BEOL metallization has changed from aluminum (Al) to Cu metallization due to performance requirements. Unlike Al, Cu bond pads are not directly bondable with gold wires during packaging. This requires an Al, NiP/Au or NiP/Pd capping. Electroless plating bath consists of metal complex, reducing agent, and one or more stabilizers. This paper discusses plating defects, as characterized by step or stair deposition, caused by electroless process which has two stabilizers. These defects are previously seen to correlate with higher stabilizer concentrations (high stability index) [5,6,7]. In this paper, we show that the lower concentration of second stabilizer (stabilizer B) can lead to step defects even if the other stabilizer (stabilizer A) concentration is on target.

II. BOND PAD METALLIZATION

The main reason, copper has replaced aluminum for BEOL metallization is because of its higher conductivity. However, the use of copper has introduced additional challenges. For example, copper is prone to corrosion and does not bond with Au. One option to overcome this is to selectively deposit NiP-Pd on copper. Electroless Pd film works as corrosion barrier for copper pads as well as provides a bondable surface for gold wire bonding. Using palladium alone is not compatible with copper, as copper diffuses into Pd at a significant rate, hence a diffusion barrier, like nickel, is needed before depositing Pd [8]. In addition, NiP film also works as a nucleation seed layer for palladium during its electroless deposition thereby avoiding copper corrosion in the palladium bath.

Figure 1 shows the schematic of the bond pad scheme. Copper bond pads are patterned, followed by dielectric films deposition and patterning, to expose on only copper pads. NiP film is selectively deposited on the exposed copper surface. A thin layer of Pd is then deposited electroless on top of the NiP film.

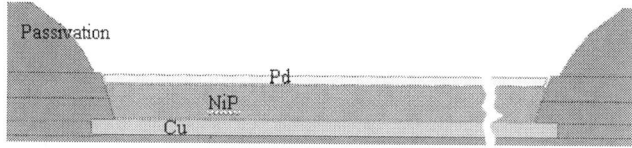

Figure 1. Copper bond pad capping by electroless NiP and Pd films (not to scale).

Electroless nickel and palladium chemistries are provided by vendors. These chemistries are plumbed to an automated batch-plating tool, which processes the wafers in the following sequence of steps:

1. **Activation**: Bond pads are activated in a Pd based catalyst. This causes adsorption of Pd on the Cu surface. Pd acts as a catalyst for electroless reactions.

2. **Rinse**: DIW rinse at room temperature to rinse off the chemicals.
3. **Nickel bath**: Wafers are submersed in electroless solution.
4. **Rinse**: DIW rinse at room temperature to rinse off the chemicals.
5. **Palladium bath**: Wafers are submersed in a Pd bath
6. **Rinse**: DIW rinse at room temperature to rinse off the chemicals.
7. **IPA vapor dry**: Wafers are dried in an IPA vapor dryer.

III. STEP DEPOSITION

Step deposition is characterized by a gradual decrease in NiP thickness toward the edge of pad. Figure 2 shows the comparison between normal deposition and step deposition.

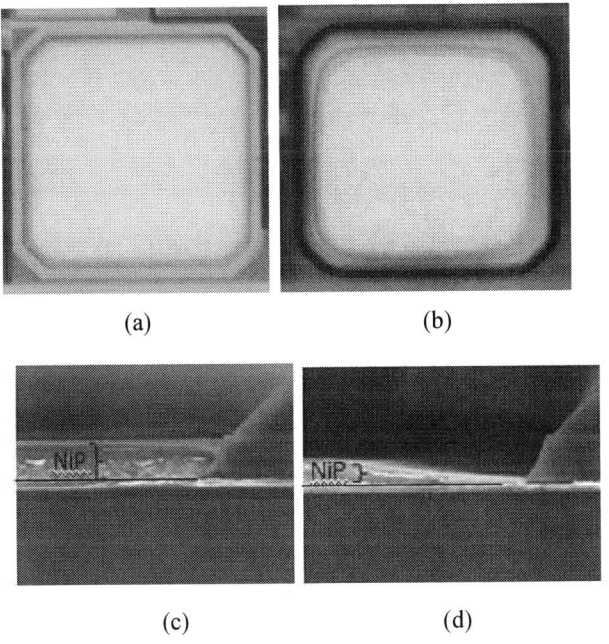

Figure 2. Bond pad images: (a) a microscope image of normal pad, (b) a microscope image of step deposition pad (c) a cross-section of normal pad, and (d) a cross-section of step deposition pad.

Step deposition is the result of slower deposition at the edge of the bond pad compared to the center of the bond pad, as shown in Figure 2d. Many factors can cause a slower deposition rate, including pH, temperature, concentrations of bath constituents, and stabilizers. Step deposition is caused by a non-linear diffusion of the stabilizer at high concentrations [5,6,7]. There are also many contaminants that can reduce the deposition rate of Ni [9,10,11]. If these contaminants are present at quite low levels in a location where mass transport of these materials to the pad surface is happening by diffusion, the same effect might be possible.

However, in this case, step deposition was not caused by higher stabilizer concentration (as the stability index was well below the upper control limit) but rather by lower concentration of stabilizer B in presence of stabilizer A.

IV. EXPERIMENTAL

The electroless nickel bath is prepared at the point of use from vendor-supplied components. These components were stored in glass beakers or plastic bottles for test purposes. For each set of experiments, a fresh bath was prepared. 400ml glass beakers were used for beaker experiments. The temperature in the beakers was maintained using a hot plate from VWR (model VMS-C7).

300mm patterned wafers with Cu bond pads broken into small pieces (~1 inch x3 inches) were used for the beaker tests. Microscope images and SEM cross-sections were used to characterize the deposition.

To characterize the bath stabilizers, an indirect method; called stability index (SI) titration was used. In this method, a bath sample is titrated with Pd activation solution. The end point is achieved when the bath decomposes. The amount of Pd solution used in titration is translated into SI. In this case, the SI is the volume of Pd solution required to reach the end point for one liter of bath. A higher SI means better bath stability (or higher stabilizer concentration).

V. RESULTS

To decipher the step deposition problem from low stability index baths, the experiments were partitioned to address three categories of potential root causes:

1. Contamination in the plating tool
2. Incoming chemical issue
3. Process margin issue

1. **Contamination in the plating tool**: Step deposition was occurring during the fresh bath qualification. Before new bath is prepared, tank is stripped of nickel deposit by HNO_3/HCl chemistry. It is a known fact that nitrates and nitrites residue adversely impacts deposition. This possibility was eliminated by the following test, where the tank was filled with Ni bath and circulated for 11hours and refilled with new bath. This eliminates any potential HNO_3/HCl contamination. However, there was no toggle for step deposition. Next, we analyzed water samples after HNO_3/HCl strip, which showed normal level of trace impurities and residues. This eliminated contamination as root cause.

2. **Incoming chemical issue**: To eliminate tool as a factor and check whether incoming chemicals have issues, tests were conducted in a beaker. However, to validate beaker approach, samples were taken from the tool bath which was defect free and tested in the beaker. The results showed that the beaker and tool were consistently good (Figure 3a). On the other hand, when the tool had step deposition issue, beaker tests also showed the same issues (Figure 3b).

The bath prepared in the beaker from components did not show the defects (Figure 3c). This test was repeated after two days of storing the components in propylene bottles; step deposition appeared (Figure 3d). Similar tests were conducted by storing the components in PTFE bottles, with no defects appearing.

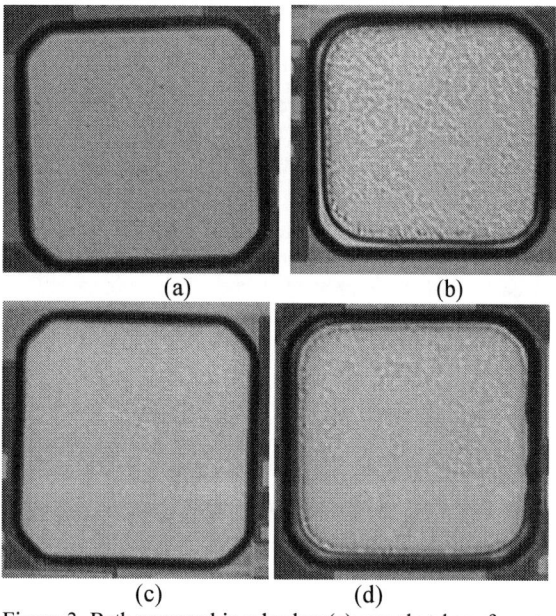

(a) (b)

(c) (d)

Figure 3. Bath prepared in a beaker (a) sample taken from tool at 71 hr of bath life, (b) sample taken from tool at 2hr of bath life, (c) bath prepared from components taken from the tool incoming source in a beaker (d) bath prepared from components taken from the tool incoming source, but stored in polypropylene bottles.

3. **Process margin issue**: The above beaker tests showed conflicting results, implying that the issue is caused by process margin. Results were dependent on the type of bottle used for storage, which is the result of adsorption of stabilizer B on the bottle surface. This was confirmed by a stability index test, as well as by the addition of stabilizer B on the bath with a defect issue (Figure 4).

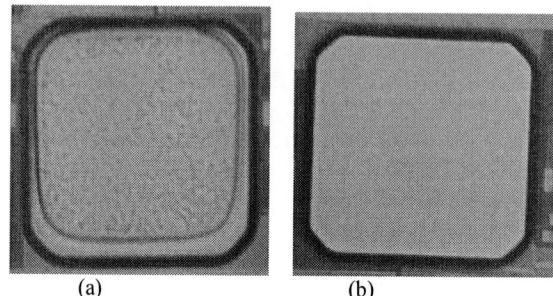

(a) (b)

Figure 4. Step deposition toggle with stabilizer B: (a) deposition from the bath with step deposition (b) deposition after the addition of stabilizer B.

VI. DISCUSSION

Electroless baths work by mixing a reducing agent with a complexed metal ion. By design, the chemical activities of the reducing agent and metal ions are such that the reaction is spontaneous (negative Gibbs free energy). However, process conditions are kept in such a way that a catalyst is needed to overcome the activation barrier. To make this activation barrier more effective, stabilizers are added to inhibit the effect of any small particles/surfaces that can work as nucleation sites. The main consideration for stabilizers is to lengthen the bath life. A higher concentration of stabilizers causes the metal deposition rate to drop, and eventually, at high enough concentrations, it stops the deposition completely, as the stabilizer blocks the active deposition sites.

For bulk surfaces, stabilizers have much wider windows. As feature sizes decrease, the stabilizer concentration window becomes narrower.

Based on the previous tests, the root cause of the step deposition is a lower stabilizer B concentration. Some stabilizers function not only as a stabilizer, but also work as accelerators [9]. In this case, stabilizer B works as an accelerator as well as a stabilizer. Based on these tests and taking into account the accelerating effect, the step deposition issue can be explained as follows:

- Bath formulation relies on the accelerating effect of stabilizer B.
- In the absence of or lower concentration of stabilizer B, stabilizer A becomes dominant.

As a bath uses accelerating effect of stabilizer B, bath activity becomes lower in the absence of stabilizer B. This leads to step deposition due to stabilizer A. If there is no stabilizer A in the bath, the lower concentration of stabilizer B should not cause step deposition.

The following tests were performed to confirm the stabilizer B behavior while keeping the stabilizer A concentration at target. Stabilizer B was added in small increments to a bath with the step defect issue. After the first addition, the deposit became smooth and defect-free. In subsequent additions, step deposition appeared, and, at a high enough concentration, there was no deposition (Figure 5).

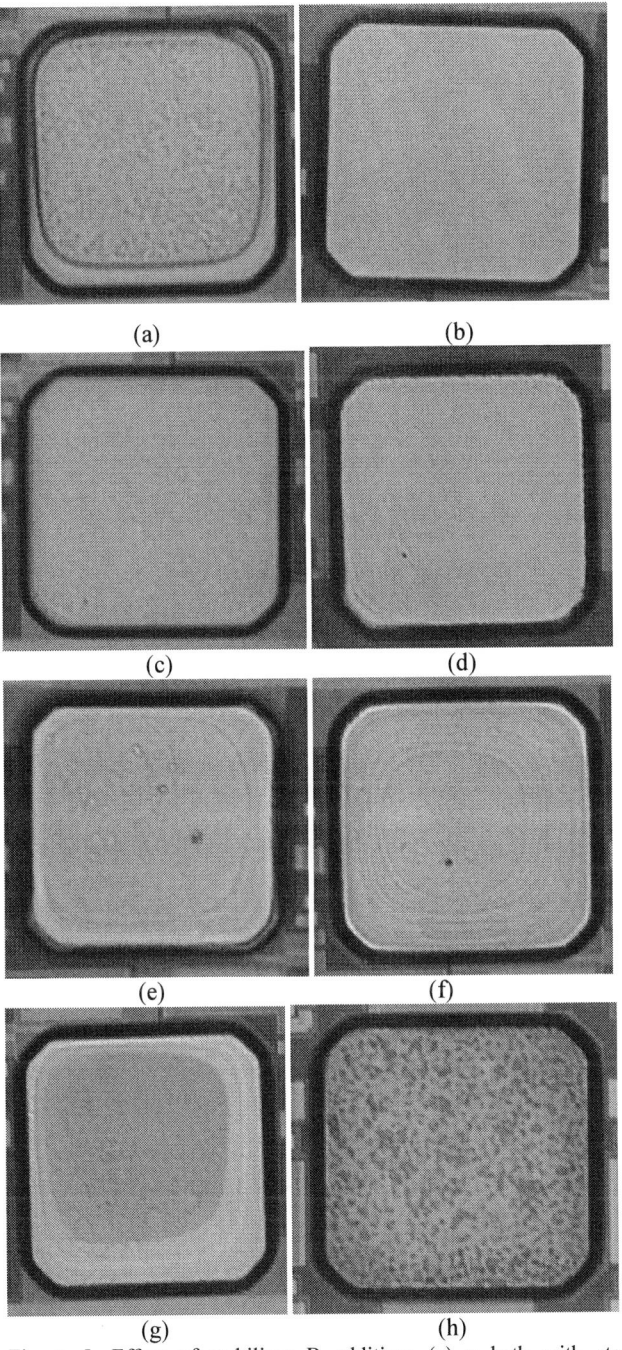

(a)

(b)

(c)

(d)

(e)

(f)

(g)

(h)

Figure 5. Effect of stabilizer B addition: (a) a bath with step deposition (b) addition of x ml of stabilizer B (c) addition of 2x ml of stabilizer B (d) addition of 3x ml of stabilizer B (e) addition of 4x ml of stabilizer B (f) addition of 5x ml of stabilizer B (g) addition of 6x ml of stabilizer B (h) addition of 7x ml of stabilizer B.

At a lower concentration of stabilizer B (Figure 5a), step deposition is caused by the dominance of stabilizer A. As stabilizer B is increased, step deposition (Figure 5c-g) and skip plating (Figure 5h) is caused by stabilizer B.

VII. CONCLUSION

Stabilizer concentrations have a narrow range of operation for bond pads. Higher concentrations of stabilizers cause step deposition and skip plating. This paper has shown that lower concentrations of one stabilizer can also lead to step deposition. Thus to operate in a desired concentration of stabilizers, it is important to formulate the bath taking into consideration the loss of stabilizer(s) by adsorption in the storage containers as well as in the plating tool.

It is common to use a stability index titration to monitor the stabilizers in electroless baths. However, this paper shows that this titration is not sufficient to reflect the interactions of multiple stabilizers and that this method needs to be supplemented with a secondary analysis.

ACKNOWLEDGMENT

We would like to express sincere thanks to David Moseman and John Verdadero for helping with our tests. Also, we would like to thank the Micron YE lab for cross-sections and the Micron Corporate lab for chemical sample analysis.

REFERENCES

[1] C. K. Hu, L. Gignac, R. Rosenberg, E. Liniger, J. Rubino, and C. Sambucetti, "Reduced Electromigration of Cu Wires by Surface Coating," Applied Physics Letters, 81 (10) 2002, 1782-1784.

[2] C. H. Lee, S. C. Lee, and J. J. Kim, "Improvement of Electrolessly Gap-Filled Cu Using 2, 2' –Dipyridyl and Bis-(3-sulfopropyl)-disulfide (SPS)," Electrochemical and Solid-State Letters, 8 (8), 2005, C119-C113.

[3] N. Shaigan, S. N. Ashrafizadeh, M. S. H. Bafghi, and S. Rastegari, "Elimination of the Corrosion of Ni-P Substrates during Electroless Gold Plating," Journal of the Electrochemical Society, 152(4), 2005, C173-C178.

[4] D. A. Hutt, C. Liu, P. P. Convay, D. C. Walley, and S. H. Mannan, "Electroless Nickel Bumping of Aluminum Bondpads-PartII: Electroless Nickel Plating," IEE Transactions on Components and Packaging Technologies, 25 (1), 2002, 98-105.

[5] Andre M. T. van der Putten and Jan-Willem G. de Bakker, "Anisotropic Deposition of Electroless Nickel," Journal of the Electrochemical Society, 140(8), 1993, 2229-2235.

[6] Kwang-Lung Lin and Chih-Hua Wu, "Structural Evolution of Electroless Nickel Bump," Journal of the Electrochemical Society, 150(5), 2003, C273-C276.

[7] S. Zhang, J. De Baets, M. Vereeken, A. Vervaet, and A. Van Castler, "Stabilizer Concentration and Local Environment: Their Effects on Electroless Nickel plating of PCB Micropads," Journal of the Electrochemical Society, 146(8), 1999, 2870-2875.

[8] I. J. Jeon, J. H. Hong, and Y. P. Lee, "Study of Interdiffusion in Pd/Cu Multilayered Films by Auger Depth Profiling," J. Appl. Phys. 75(12), 1994, 7825-7828.

[9] Glenn O. Mallory and Juan B. Hajdu, Electroless Plating: Fundamentals and Applications. New York: Noyes Publications, 1990, pp 34-39, 107-109

[10] Frank Altmayer, "P2 for Electroless Nickel," Plating & Surface finishing, November 2000, 29-31.

[11] Gerald Laitinen, "Troubleshooting Electroless Nickel," Plating and Surface Finishing, 89(1), 2002, 8-12.

ON-CHIP 3D INDUCTORS USING THRU-WAFER VIAS

Gary VanAckern, R. Jacob Baker, Amy J. Moll, Vishal Saxena
College of Engineering
Boise State University
1910 University Dr. Boise, ID 83725

Abstract—**Three-dimensional (3D) inductors using high aspect ratio (10:1); thru-wafer via (TWV) technology in a complementary metal oxide semiconductor (CMOS) process have been designed, fabricated, and measured. The inductors were designed using 500 μm tall vias, with the number of turns ranging from 1 to 20 in both wide and narrow-trace width-to-space ratios. Radio frequency characterization was studied with emphasis upon de-embedding techniques and the resulting effects. The open, short, thru de-embedding (OSTD) technique was used to measure all devices. The highest quality factor (Q) measured was 11.25 at 798 MHz for a 1-turn device with a self-resonant frequency (f_{sr}) of 4.4 GHz. The largest inductance (L) measured was 45 nH on a 20-turn, wide-trace device with a maximum Q of 4.25 at 732 MHz. A 40% reduction in area is achieved by exploiting the TWV technology when compared to planar devices. This technology shows promising results with further development and optimization.**

Keywords- 3D; Integrated Inductors; Thru-Wafer Vias (TWVs), Thru-Silicon Vias (TSVs);

I. INTRODUCTION

The interest and proliferation of radio-frequency (RF) circuits in recent years has provided broad opportunity for the development of front-end RF modules, such as the voltage-controlled oscillators (VCOs), low-noise amplifiers (LNAs) and switching regulators needed to support new wireless applications [1, 2], including multi-mode wireless technology [3]. These RF modules have their foundation built upon discrete passive circuit components like the high frequency (HF) inductor. In the last decade, integration of monolithic inductors in silicon-based complementary metal oxide semiconductors (CMOS) has been realized and is preferable, due to the aggressive scaling in MOS devices, and its improved performance above 1 GHz [4, 5]. This movement has the added benefit that does not rely on off-chip components [3].

As devices scale, designers are challenged with producing smaller and more efficient RF building blocks, while maintaining or improving circuit performance, predictability, and robustness [2]. These three design requirements directly influence the selection of passive components used in the building blocks and thus have fueled the quest for an improved integrated inductor.

While the inductive coil has been around for over 100 years, its wide-spread use in modern CMOS circuits has been limited by its relatively large size (when compared to other

circuit elements) and its inherent performance and integration limitations.

In order to achieve a reasonable inductance value (~10 nH), a planer inductor needs to be designed and manufactured with an extremely large footprint, on the order of 250 um². Unfortunately, increasing the inductor size both increases the manufacturing cost and produces undesired parasitic effects thus reducing its fundamental performance factors. This includes a poor quality factor (Q), a reduction in self-resonant frequency (f_{sr}), and a low inductance value (L). With this information in hand, circuit designers will be able to optimize and further experiment with new design solutions to achieve a better integrated inductor. This paper provides a starting point for an alternative inductor design, a 3D inductor using through-wafer vias (TWVs), also known as through-silicon vias (TSVs). An example layout/cross-section is seen in Fig. 1. This image of the 3D inductor was generated, and ultimately simulated, with Ansoft's HFSS software.

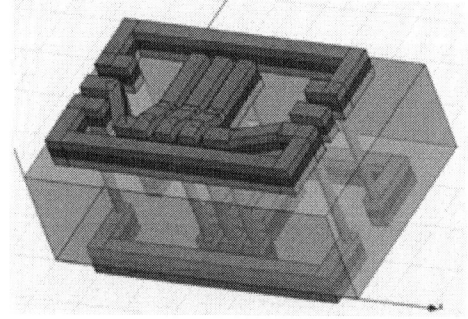

Figure 1. HFSS 3D Inductor Architecture

II. INDUCTOR PHYSICS

A. The Inductive Phenomena

An A/C current flowing in a straight wire, a simple coil of wound wire (solenoid), or a CMOS monolithic planar spiral inductor gives rise to the magnetic field intensity H, measured in units of A/m, and is related to the magnetic flux density B, measured in units of Tesla, as seen below in (1). The magnetic permeability μ is absolute magnetic permeability.

$$B = \mu H \tag{1}$$

The total magnetic flux is equal to the integral of the magnetic flux density over an area of a surface S that intersects the field lines. In the special case of a planar surface, this can be simplified where A is the cross-sectional area of the intersecting surface and θ is the angle between the surface and the magnetic field lines that extend normal to the flow of current. In other words

$$\Phi_m = \int_S B \cdot dS \;\rightarrow\; \Phi_m = BA\cos\theta \qquad (2)$$

Flux linkage (λ) represents the total magnetic flux passing through a surface S of a single loop of current-carrying wire. This is covered in (3), where N is the number of loops. For example, if two N-turn loops are tightly wound around S, the magnetic flux generated from each loop is shared through both loops. As such, the total magnetic flux linkage is increased by the square of the number of loops N times the magnetic flux of one loop of wire as shown below. The quantity of inductance can be determined by the ratio of the flux linkages to the current that creates the magnetic flux, as shown in (4). In other words, inductance is primarily a function of geometric shape.

$$\lambda = \Phi_T \cdot N \;\rightarrow\; \lambda = \Phi_1 \cdot N^2 \qquad (3)$$

$$L = \frac{\lambda}{I} = \frac{\phi_T \cdot N}{I} = \frac{\mu \cdot N^2 \cdot \pi \cdot a^2}{h} \qquad (4)$$

B. Mutual- and Self-Inductance

Using the method presented in the last section, two types of inductance make up the total inductance: mutual- and self-inductance. Mutual-inductance is a result of the proximity effect occurring between two closely spaced circuits, circuit elements, or wires [7] in series or parallel. Accordingly, this depends on the amount of flux linkages interacting between the two elements. Illustrated in Fig. 2 are two 3-loop coils of wire with interacting flux linkages. Coil A is being driven by current I_A, and as such is creating the flux density from coil A, while coil B is not being driven, but rather receiving.

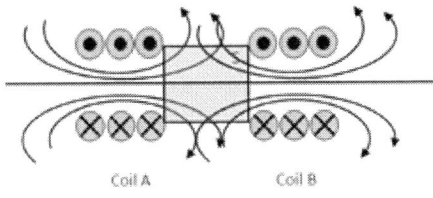

Figure 2.   Mutual-Inductance of Two Coils

C. Electric/Magnetic Fields

The planar square spiral provides the best illustration of the electromagnetic fields exhibited and the utility of a physical model. As illustrated in Fig. 3, one magnetic and three electric fields are produced when an AC voltage is applied to port 1 [6]. The effects of the electric fields are modeled using capacitors while magnetic fields are modeled by inductors. Figure 4 shows the lumped circuit-equivalent planar inductor circuit model.

The first electric field E_1 is a result of the voltage difference between the terminal connections of the spiral and is simply due to ohmic losses in the traces [6]. This is directly dependent upon material resistivity (ρ) and is modeled as the series resistance R_S. The second electric field E_2 is a consequence of the voltage difference between any two turns in the spiral and any individual turn and the underpass [6]. This is a consequence of the second port being connected using a lower level of metal, which induces an inter-winding parasitic capacitance due to the presence of the interlayer dielectric. The modeling parameter for E_2 is C_P [6]. The third electric field E_3 is present due to the voltage difference between the silicon substrate and the metal of the spirals.

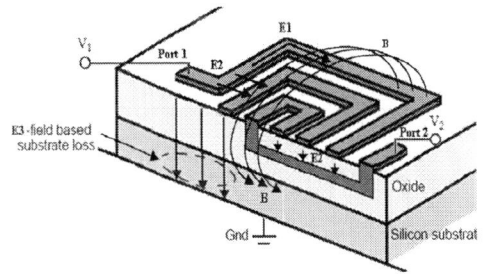

Figure 3.   Electric and Magnetic Fields in a CMOS Planner Inductor

Field E_3 induces capacitive coupling to the substrate and is often times the most predominant parasitic since it extends into the substrate [6]. This is modeled as the parameter C_{OX}. The effect of this field is made worse because many CMOS circuits use low-resistivity substrates, an epitaxial layer, having a resistivity in the range of < 10 Ω/cm. This allows for current to flow in the substrate easily. Due to this current flow, it is necessary to include modeling parameters for the intrinsic substrate capacitance and resistance. These parameters are identified as C_{SUB} and R_{SUB}, respectively.

The final field is the magnetic field B produced by the AC current that flows through the traces of the spiral. While the magnetic field is what induces the desired inductive behavior, it also creates a complementary parasitic behavior in the metal traces due to eddy-currents [4, 6, 7, 8, 9].

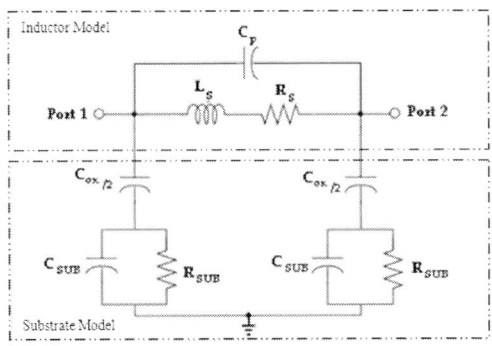

Figure 4.   Equivalent Planar Inductor Circuit Model

III. 3D INDUCTOR FABRICATION

A. Fabrication

Fig. 5 illustrates the first generation of 3D masks for a 1-turn inductor where a) shows the via mask, b) shows the top metal mask with ground-signal-ground (GSG) probe pads and guard ring, and c) shows the bottom metal mask with guard ring. Fig. 5d depicts all masks overlaid to illustrate mask alignment. The Fig. 5e micrograph shows the topside view of a fabricated device. An optical resolution limitation was identified due to the dry-film photoresist which caused subsequent turns to become shorted. In Fig. 5f a SEM of the device clearly shows this. As such, a line-to-line spacing design rule of 30 μm minimum became necessary.

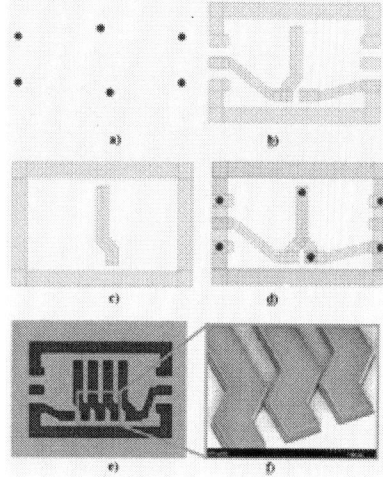

Figure 5.   1st Generation 3D Inductor Masks and Fabricated Device

Due to the new design rule, a second set of masks were needed that traded polygon lines for straight lines oriented off-angle. Mask redesign allowed for the addition of narrow-trace (NT) and wide-trace (WT) line-to-line spacing devices. Fig. 6 shows similar masking levels as Fig. 5 with the addition of these new devices.

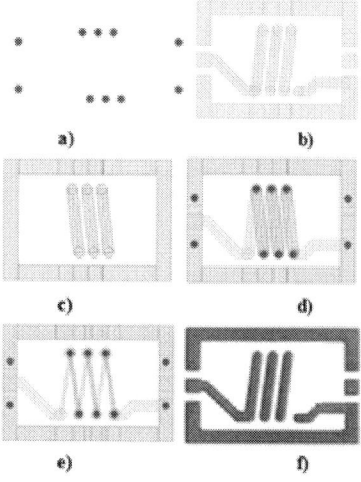

Figure 6.   2nd Generation 3D Inductor Masks and Fabricated Device

The TWV fabrication process flow is beyond the scope of this paper. The interested reader can review works reported at BSU covering solid and barrel-coated via methods.

IV. MEASUREMENT TECHNIQUE

A. Equipment Setup/Calibration/De-Embedding

The 3D inductors were characterized with an HP8510C vector network analyzer (VNA) in conjunction with a manual Cascade™ Microtech Summit microwave probe station, 150 μm pitch ground-signal-ground Infinity Probes, a standalone PC, and a non-conductive auxiliary chuck built to isolate the bottom metal traces during measurement.

The short, open, load, and thru (SOLT) calibration method [10, 11] was used to remove parasitic factors between the VNA and the probe tips with a reasonably accurate (±10%) measurement below 20 GHz. Measurement repeatability is obtained by manually placing the probe tips exactly centered on the probe pads.

V. MEASURED 3D INDUCTOR PERFORMANCE

The highest quality factor observed below the self-resonant frequency (f_{sr}) on the measured devices, occurred on the 1-turn inductors and measured Q_{max} values of 11.25 (WT) and 7.84 (NT). Increasing the number of turns was dominated by parasitic components and decreased with increasing N-turns as shown in Fig. 7.

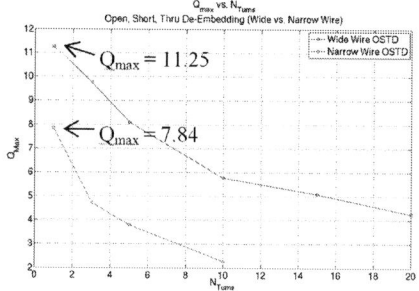

Figure 7. Q_{max} vs. N_{Turns}

The self-resonant frequency can be obtained from the quality factor plot or the impedance plot of |Z|. Fig. 8 shows this in the plot of f_{sr} versus N-turns. The NT devices resulted in a higher $f_{sr} \sim 4.58$ GHz for a 1-turn device, while the WT devices degraded at a slightly faster rate with increasing N.

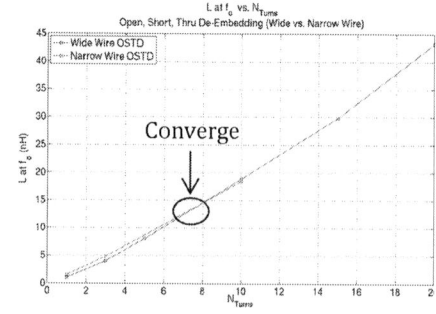

Figure 8. f_{sr} vs. N_{Turns}

978-1-4577-1735-2/12 $31.00 © 2012 IEEE

The basic 1-turn 3D inductor measures inductance values of 1 nH (WT) and 1.5 nH (NT). As expected, inductance increases with the number of turns due to the increase in flux linkages. The measured inductance at the characteristic frequency f_0, versus the number of turns between the NT and WT devices, converges at N=8 as seen in Fig. 9.

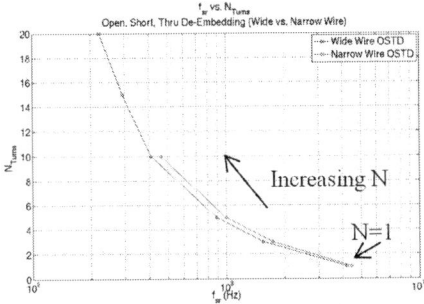

Figure 9. L vs. N_{Turns}

The WT 1-turn device measures a $|Z|$ of 4.8 Ω, while the 20-turn device measures 20Ω at f_0. The resistance increases non-linearly with an overall average increase of 2 Ω per-turn. The WT devices overall measured lower in resistance than the NT devices. The 1-turn WT inductor measured 54% less resistance than the 1-turn NT inductor. Fig. 10 illustrates the impedance growth with increasing N values.

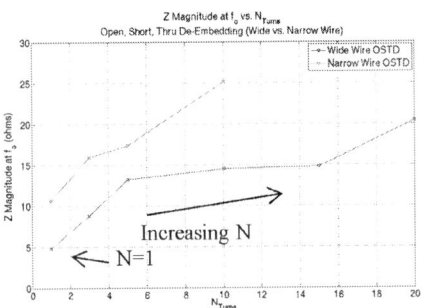

Figure 10. $|Z|$ at f_o vs. N_{Turns}

The phase angle, also measured at f_o, measured higher on the WT devices, in the range of 84° to 76° for $N = 1$ and $N = 20$, respectively. The NT devices measured angles in the range of 82° to 66° for $N = 1$ and $N = 10$, respectively. Since phase angles at 90° cause resonance, a lower phase angle provides increased margin that ensures reliable performance.

VI. CONCLUSION AND FUTURE WORK

The 3D TWV inductor architecture provides a 40% smaller device footprint when compared to an equivalent N-turn planer device. The 3D 1-turn wide inductor achieved a maximum WT device de-embedded Q of 11.25 and $f_{sr} = 4.4$ GHz. While the 1-turn WT device measured ~1 nH and increased non-linearly to ~45 nH up to 20-turns. Convergence between WT and NT devices occurs at $N = 8$ turns, with the NT device providing higher inductance below $N = 8$ and the WT device providing higher inductance above N = 8. The 1-turn WT device series resistance measured 4.8 Ω and increases to 20 Ω for $N = 20$.

However, each additional turn added drops off to 1.02 Ω per turn above $N = 15$.

As with the planar inductor, the 3D TWV inductor suffers similarly from capacitive coupling to the substrate. As such, future work on this architecture should be focused on optimization of the via height (wafer thickness), the via pitch, the inductor radius, and the line-to-width space ratio. The architecture would also benefit from devising a scheme to either remove or replace the silicon substrate within the core of the inductor.

ACKNOWLEDGMENT

This work was funded in part by Government research contract N66001-01-C-8034 to develop TWV technology and was executed in conjunction with Research Triangle Institute.

REFERENCES

[1] Chang, C.A.; Tseng, S.; Chuang, J.Y.; Jiang, S.; Yeh, J.A.;, "Characterization of spiral inductors with patterned floating structures," Microwave Theory and Techniques, IEEE Transactions on , vol.52, no.5, pp. 1375- 1381, May 2004

[2] Wu, C; Tang ,C.; Chiu, C.; Liu, S., "Analysis and application of miniature 3D inductor," *Circuits and Systems, 2002. ISCAS 2002. IEEE International Symposium on* , vol.2, no., pp. II-811- II-814 vol.2, 2002

[3] Chen, J.; Zou, J.; Liu, C.; Schutt-Aine, J.E.; Kang, S.-M.K.;, "Design and modeling of a micromachined high-Q tunable capacitor with large tuning range and a vertical planar spiral inductor," *Electron Devices, IEEE Transactions on* , vol.50, no.3, pp. 730- 739, March 2003

[4] Shi, X.; Ma, J.; Yeo, K.S.; Do, M.A.; Li, E., "Equivalent circuit model of on-wafer CMOS interconnects for RFICs," *Very Large Scale Integration (VLSI) Systems, IEEE Transactions on* , vol.13, no.9, pp. 1060- 1071, Sept. 2005

[5] Langford-Smith, F. Radio Designer's Handbook. Newnes, 1997.

[6] Aguilera, J., et al. "A guide for on-chip inductor design in a conventional CMOS process for RF applications," *Applied Microwave & Wireless, 2001.* pp.56-65, Oct. 2001

[7] Talwalkar, N. "Integrated CMOS Transmit-Receive Switch Using On-Chip Spiral Inductors," Ph.D Dissertation, Stanford University, Stanford, December 2003

[8] Krafcsik, D.M.; Dawson, D.E.;, "A Closed-Form Expression for Representing the Distributed Nature of the Spiral Inductor," *Microwave and Millimeter-Wave Monolithic Circuits* , vol.86, no.1, pp. 87- 92, Jun 1986

[9] Kodali, S.; Allstot, D.J.;, "A symmetric miniature 3D inductor," *Circuits and Systems, 2003. ISCAS '03. Proceedings of the 2003 International Symposium on* , vol.1, no., pp. I-89- I-92 vol.1, 25-28 May 2003

[10] Cascade; "A Guide to Better Vector Network Analyzer Calibrations for Probe Tip Measurements", Cascade Microtech Application Note, TECHBRIEF 4-0694, 1994

[11] Kolding, T.E.;, "On-wafer calibration techniques for giga-hertz CMOS measurements ," Microelectronic Test Structures, 1999. ICMTS 1999. Proceedings of the 1999 International Conference on , vol., no., pp.105-110, 1999

An Algorithmic Study of DDR3 SDRAM On-Die Termination Switch Timings

S. N. Wong

Test Engineering
Micron Semiconductor Asia Pte. Ltd.
Singapore
wongshihnern@micron.com

Abstract—**Given the increasing clock signaling rates used by digital memory subsystems, it has become progressively more important that DDR3 SDRAM provide well-controlled impedance to the signal transmission path. The DDR3 specification requires flexible termination, and conformance to analog timing specifications under various operating modes is necessary to ensure reliable system operation. The *direct* measurement and test of these timings is not possible using simple pass/fail voltage comparators, in the absence of purpose-built hardware (to measure the linear-extrapolated voltage slew rates as defined in the specifications sheet). To get around this limitation, these timing limits are tested in the backend component test flows with two test comparator strobes (per spec) placed at *alternative* timing and voltage positions. These strobes judge and ensure compliance to MIN and MAX (fastest and slowest) ODT transition timings. The selection of voltage-level and timing-edge offsets requires time-consuming analysis of "shmoo" charts, which are test comparator pass/fail results plotted out across different voltage and timing sweep positions. In this paper, we will share the theory behind an algorithm to *visually* recognize and extract various signal transition metrics, including slew rates, spec/test points, test-point, and voltage/time offsets from shmoo charts automatically.**

Keywords- DDR3; Tester enhancement; on-die termination;

I. INTRODUCTION

At typical operating speeds ($\geq$ 800 MHz) of current DDR3 memory systems, the use of per-component on-die termination (ODT) is essential to improving channel signal fidelity. Traditionally only required for analog systems, these termination impedances (R_{TT}) can be switched on/off and toggled among different values. These features enable the memory controller to adapt and optimize impedance matching of the memory channel to changing load conditions. Explicit switch timing parameters (e.g., t_{ADC}, t_{AON}, t_{AOF}) are specified for these functionalities in the data sheet. Consequently, the ability to test these timing specifications adequately is essential to maintaining product quality for our customers.

Functional testing of ODT relies on observing changes to voltage levels at device input/output pins as a result of changing device termination settings, as well as the tester loading condition. Appropriate tester comparator levels can then be set when observing changes to the ODT resistance settings over time. Referring to the equivalent circuit model in Fig. 1, the voltage level output V_{OUT} is given by (1):

$$V_{OUT} = V_{DDQ} \frac{R_{TERM}}{2(R_{TT} + R_{TERM})} \quad (1)$$

Theoretical V_{OUT} levels (based on allowable tolerances in the datasheet for V_{DDQ}=1.6V and R_{TERM}=60Ω) are tabulated in Table I. These theoretical values are computed under DC conditions, and we expect changes in R_{TT} to result in slow (relative to clock rates) changes to V_{OUT}. In practice, due to wide variations in R_{TT} values allowed by the spec sheet, the optimal comparator level cannot be predicted purely from (1) In addition, factors like variations in the physical characteristics of each pin channel (parasitic capacitance, pin resistance, etc.), as well as part-to-part variations, prevent the effective *prediction* of the optimal comparator level without actual "field" measurements at the tester. The determination of the optimal comparator position requires a lengthy procedure modeled after the spec sheet definition.

Shown in Fig. 2 is the definition of t_{ADC} (dynamic to nominal R_{TT} values) in the data sheet [1]. This parameter specifies the time taken for the ODT to transition from dynamic/write-R_{TT} to nominal-R_{TT}, defined as the time delta from the rising edge of CK-CK#, to the *linearly-extrapolated* point at V_{RTT_WR}. Measurement of this timing parameter *directly* is not possible with a single comparator (pass/fail) test-strobe. If we now sweep an "expect-high" comparator strobe across both timing and voltage levels within the shaded bounding box highlighted in Fig. 2, and then plot out the pass/fail results of *each* test-comparator strobe (pass/fail points are indicated by * and - respectively), we obtain the shmoo plot shown in Fig. 3. It becomes clear we can equivalently test $t_{ADC(MAX)}$ (slowest transition) at TP (Test-point) instead of at SP (Spec-point), using a single compare. We can similarly test for $t_{ADC(MIN)}$ (fastest transition) by repeating the shmoo with the test comparator inverted to "expect-low" and then performing the same steps above. This laborious procedure needs to be repeated across multiple ODT spec timings and over multiple parameters like supply voltage, latency, and termination values.

The principal goal of this study is an automated system to interpret shmoo plots *visually*, allowing us to extract characterization values across multiple parts/pins quickly:
1. Test point offsets in time (EPX) and voltage (TPY)
2. Slew rate

These characterization values allow us to more effectively/consistently quantify and compare between

978-1-4577-1735-2/12 $31.00 © 2012 IEEE

different device generations and design mask revisions, over a larger volume representation. The challenge was to develop an algorithm that can extract these parameters robustly, across different transition types, while unaffected by random variations.

II. THEORY

A. Edge detection and noise reduction

At the initial stage, we detect the signal transition/edge by searching over each column for the *first* change from pass to fail point. Referring to Fig. 4, the transition/edge points identified by the above rule are indicated by red 'plus' markers. We can view these as digitally sampled signal points s_n over time. As we shall see at the next step, high-frequency noise, which arises from repeated test pattern runs, will need to be removed from this signal, while minimizing phase distortion in the time-domain. The M-point 2-pass moving average (finite impulse response or FIR) digital filter is optimal for this problem [2], with a frequency response defined by (2):

$$H[f] = \left[\frac{\sin(\pi f M)}{M \sin(\pi f)} \right]^2 \quad (2)$$

In the frequency domain, slow roll-off and poor stopband attenuation are expected, with the trade-off being good performance (feature detection) in the time domain. The strategy for determining the optimal window size is simply to examine the spectral content of the raw edge points in the frequency domain. Taking a 1024-point Fast Fourier Transform (FFT) of the signal in Fig. 3, we see most of the signal energy in the low-frequency region, as expected. (See Fig. 5, dashed and solid lines for pre- and post-filtered signals respectively)

The (normalized) frequency f_{3dB} at which the signal energy drops to half (–3dB) for this example is 0.005. The window size that is chosen is given by the *largest* M (M_{OPT}=7 for this case), which attenuates not more than 0.5% at f_{3dB} of the raw signal, i.e. satisfying (3):

$$H[f_{3dB}]\big|_{M=M_{OPT}} \geq 0.995 \quad (3)$$

Looking at the time-domain plot in Fig. 4, we see effective removal of white noise; with minimal phase distortion. (blue 'circle' markers denote post-filtered signal points.)

B. Differentiation and obtaining pin statistics

Next, we numerically differentiate the filtered digital signal u_n using the method of central (finite) differences [3] by (4):

$$d_n = \sum_{k=1}^{(N-1)/2} c_k \left(u_{n+k} - u_{n-k} \right) \quad (4)$$

where (for N=7):

$$c_1 = +\tfrac{45}{60}, \ c_2 = -\tfrac{9}{60}, \ c_3 = +\tfrac{1}{60} \quad (5)$$

TABLE I. V_{OUT} LEVELS FOR FUNCTIONAL TEST OF ODT

V_{OUT}	Termination condition		
	R_{TT}	$R_{TT(MIN)}$	$R_{TT(MAX)}$
522-615mV	20 Ω	18 Ω	32 Ω
444-552mV	30 Ω	27 Ω	48 Ω
387-500mV	40 Ω	36 Ω	64 Ω
308-421mV	60 Ω	54 Ω	96 Ω
190-286mV	120 Ω	108 Ω	192 Ω

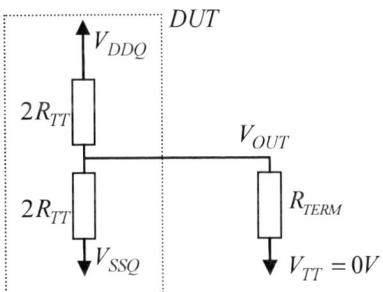

Figure 1. Equivalent circuit for ODT functional test

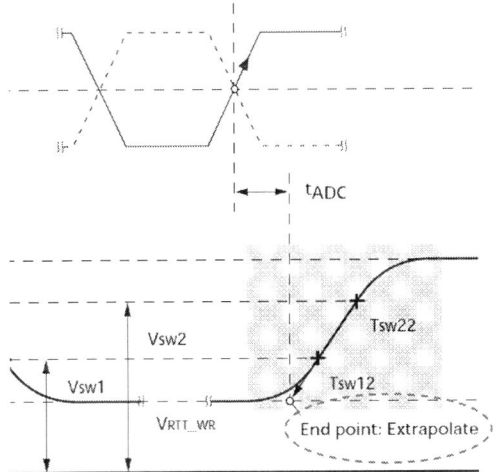

Figure 2. Definition of t_{ADC} for dynamic-to-nominal R_{TT} transition

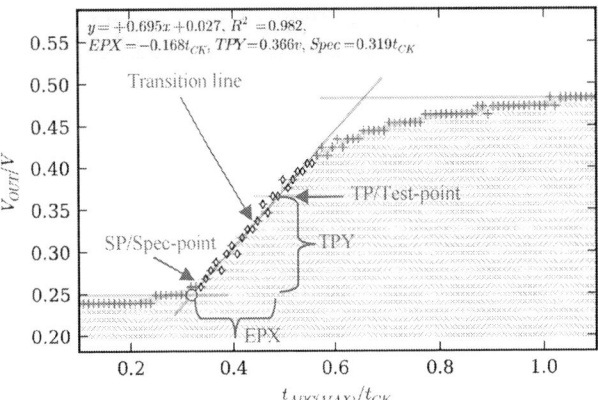

Figure 3. Dynamic ODT (dynamic-to-nominal R_{TT}) $t_{ADC(MAX)}$

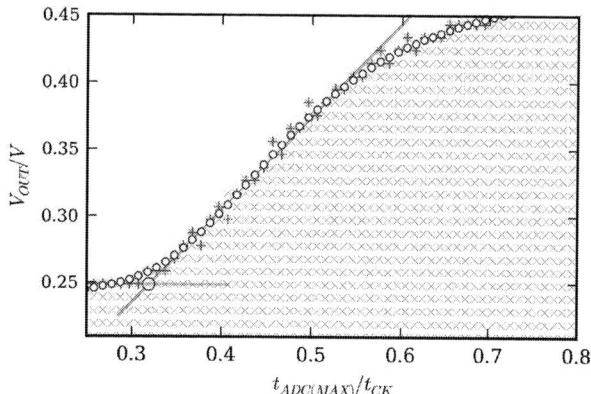

Figure 4.   Filtered signal

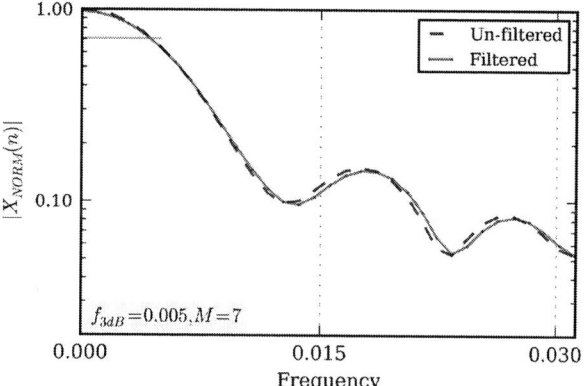

Figure 5.   1024-point FFT of signal

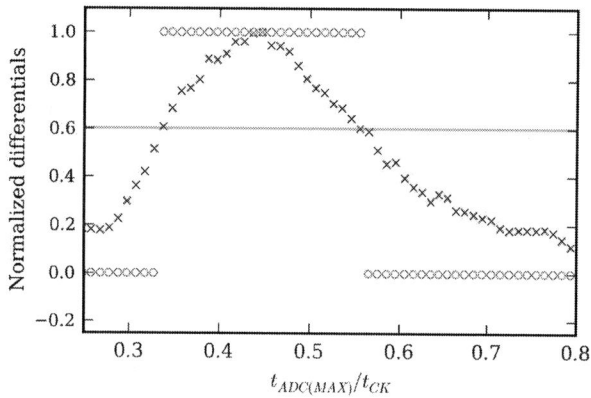

Figure 6.   Numerical differentials (normalized) of filtered signal

This approximation gives reasonably good results for the current problem, as shown in Fig. 6, with differentials marked by "X" and normalized to maximum, and with the numerical differentials peaking *near* the linear portion of the transition region, as desired. The next step of the algorithm applies a dynamic-threshold function on the differentials after normalizing to the peak or minimum of the differentials for rising and falling transitions, respectively, to identify transition points (green 'diamond' markers in Fig. 3). The optimal threshold for each pin/shmoo-plot is determined

iteratively/dynamically, by starting off with a ratio of 0.6 and progressively increasing in 0.05 steps until we see no further improvement to R^2. R^2 is the "goodness of fit" metric for the best-fit straight line through the transition points. Finally, we scale all numbers appropriately to the time and voltage units, respectively, and obtain spec-point (SP) as the intersection of the transition line with the initial horizontal/voltage level of the shmoo plot. Together with an automatically chosen TP at a mid-level V_{OUT}, TPY and EPX can then be computed as defined previously.

III.   RESULTS

The algorithm described in the previous section was tested on 239 DDR3 two-Gigabit x8 parts and was repeated four times for a total of 900 pins for multiple ODT specs: Dynamic ODT (between 120Ω R_{TT_WR} and 20Ω R_{TT_NOM}) on/off, Dynamic ODT (60Ω) on/off, and Synchronous ODT on/off (20/30/40/60/120Ω). See Fig. 9 for an example of Synchronous ODT (20Ω) $t_{AOF(MAX)}$ measurement to verify the applicability of the algorithm to the various conditions. These tests are a good representation of typical transitions such as we expect to encounter in the test flow, such as rise/fall, with/without overshoot, across multiple voltage/impedance levels.

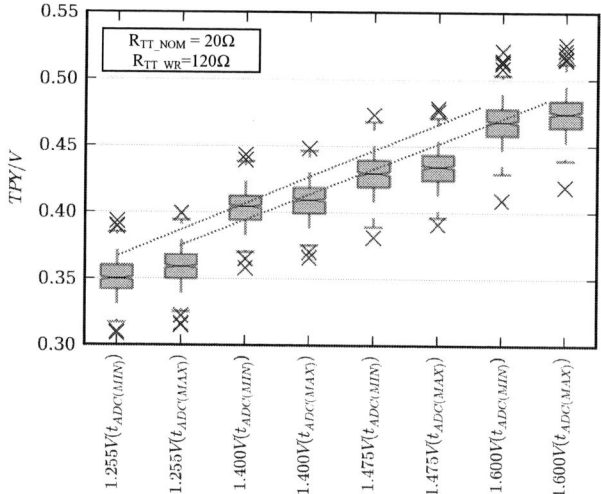

Figure 7.   TPY (V) vs. V_{DD} for Dynamic ODT (Nominal-R_{TT} to write-R_{TT})

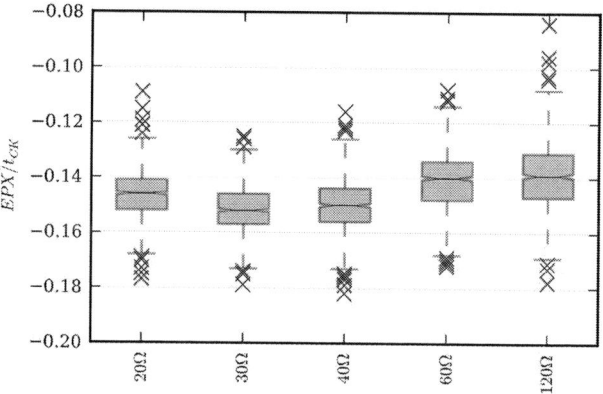

Figure 8.   EPX (t_{CK}) vs R_{TT} for Synchronous ODT $t_{AOF(MAX)}$

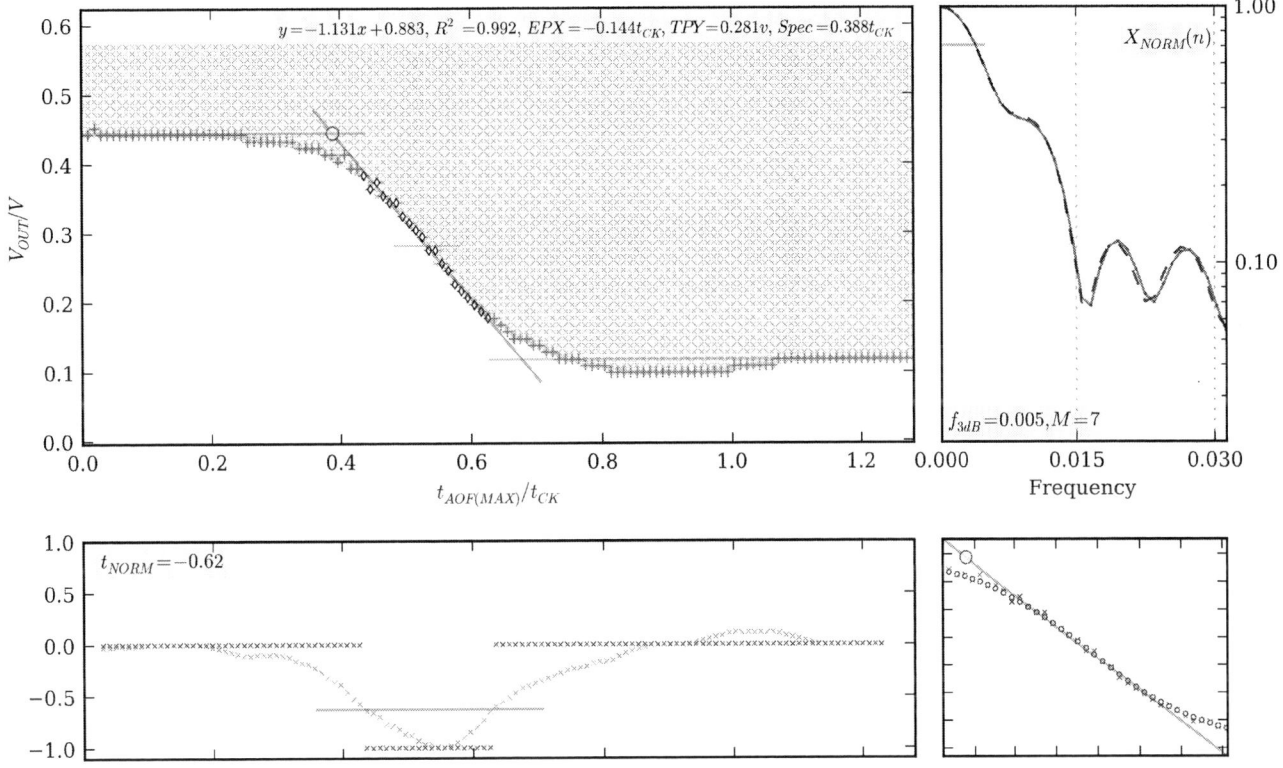

Figure 9.   Example of Synchronous ODT (20Ω) $t_{AOF(MAX)}$ measurement

A.   Varying supply voltages

For the Dynamic ODT (20Ω R_{TT_NOM} to 120Ω R_{TT_WR} transition) set of tests, multiple variants with various supply voltages are defined in the short/long test flows. The corresponding TPY levels, determined automatically from the automated algorithm described in this paper, are shown in Fig. 7. The linear relationship observed is further exploited by the test program to simplify test condition definitions.

B.   Varying R_{TT} (impedance) values

For the Synchronous ODT group of tests (turning off termination) over multiple R_{TT} values, the results in Fig. 8 indicate a single EPX value can be used for all five R_{TT} values even though TPY and slew rates vary.

IV.   CONCLUSIONS

The proper timing operation of ODT is an important enabler for higher data rates in today's large, multi-drop memory subsystem containing multiple DRAM modules. Testing these timing specifications correctly requires lengthy offline/manual engineering work to determine optimal alternative test-point comparator offsets (for the single-shot production flow tests) from multiple shmoo plots.

Through a time/frequency study of the shmoo plots, we were able to design an adaptive algorithm, which adjusts filter parameters to the measured spectral content. It then applies adaptive thresholds to the computed differentials of each

shmoo plot in order to identify the linear portion of the signal transition. The resulting code implementation allowed us to automatically interpret shmoo plots visually and extract essential pin statistics (EPX/TPY/slew rate) from large numbers of shmoo plots easily and consistently. These compare favorably with values obtained from the traditional manual method.

This development allowed us to methodically determine optimal strobe levels/timings for the ODT test registers quickly and cost-effectively, with minimal manual analysis. Additionally, this new slew rate measurement capability offers fresh insights and quantification into part capability across different fab masks and device generations, further enhancing the tester capabilities of an already great tester platform.

V.   ACKNOWLEDGMENT

The author would like to thank all colleagues and supervisors at Micron-Boise and Micron-Singapore for their ideas and suggestions.

VI.   REFERENCES

[1]   Micron Semiconductor Inc., "2GB DDR3 SDRAM datasheet", Rev. L 08/10 EN.

[2]   S. W. Smith, *The Scientist and Engineer's Guide to Digital Signal Processing* (California Technical Publishing, 1997).

[3]   B. FornBerg, Generation of Finite Difference Formulas on Arbitrarily Spaced Grids, *Mathematics of Computation*, 1988, 699-706.

Multi-bit Continuous-time Delta-Sigma Modulator for Audio Application

Rajaram Mohan Roy Koppula, Sakkarapani Balagopal and Vishal Saxena

Electrical and Computer Engineering Department, Boise State University
Boise, ID 83725-2075
Email:rajaramkoppula@u.boisestate.edu

Abstract— The design considerations for low-power continuous time (CT) delta-sigma ($\Delta\Sigma$) modulators is studied and circuit design details for a 13.5 bit modulator are given. The converter has been designed in a 0.5 um C5FN AMI CMOS technology and achieves a maximum signal-to-noise ratio (SNR) of 85 dB in a 48 kHz bandwidth and dissipates 5.4 mW from a 5 V supply when clocked at 6.144 MHz. It features a third-order active-RC loop filter, a 4-bit flash quantizer along with a Data Weighted averaging (DWA). The loop filter architecture and its coefficients have been targeted for the minimum power dissipation. The DWA also has been implemented by standard cell based synthesis to further optimize power. The figure of merit (FoM) of the CT-$\Delta\Sigma$ modulator is 3.71 pJ/bit. The fabricated chip of the modulator occupies an area of 4.5 mm².

Index Terms—Analog to digital converter (ADC), Digital to analog converter (DAC), delta-sigma modulation, noise-shaping.

I. INTRODUCTION

In recent years, due to an increasing demand for high resolution, low band width data converter systems for digital audio and voice applications (see [1] and references therein). The delta-sigma ($\Delta\Sigma$) converters [2] are well suited because of the over sampling rate (OSR), hence useful for low frequency and use noise shaping to achieve high resolution [3]. More specifically, a CT-$\Delta\Sigma$ modulator [4] [5] is preferred due to the advantages like low power consumption, inherent anti-aliasing, fixed resistive impedance and relaxed performance of the integrators [3].

A CT-$\Delta\Sigma$ modulator is characterized by the number of levels or number of bits of the internal quantizer. A single bit CT-$\Delta\Sigma$ modulator is highly linear. But, as the order of the modulator gets higher i.e., the higher order noise transfer function (NTF) suffers from the signal dependent stability limitation. And, as out-of-band gain (OBG) increases the variance of noise gets higher, the quantizer overloads more often. Further, the maximum stable amplitude (MSA) falls rapidly. However a multi-bit CT-$\Delta\Sigma$ modulator enhances the stability by impressive margins for a third or higher order NTF. With a multi-bit modulator the least significant bit (LSB) size reduces, and hence lower the probability of the quantizer overload [3] and thus enhancing the stability of the system. And the quantization noise power at the output is reduced significantly and the signal-to-noise-plus-distortion ratio (SNDR) increases at the rate of 6 dB per bit of the internal quantizer used [2]. And from the system point of view a multi-bit modulator would relax the constraints on the ensuing decimation filter [6]. Figure 1 shows the block diagram of CT-$\Delta\Sigma$ modulator.

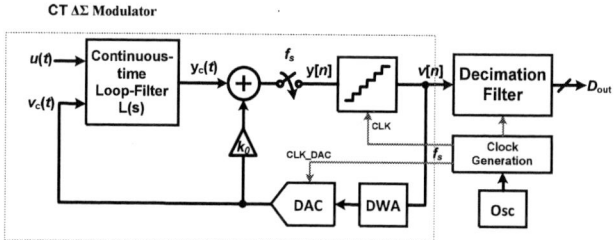

Figure 1 Block diagram of a Continuous-time Delta-Sigma ADC Modulator.

However, the dynamic range improvement of a multi bit quantizer is not quite realized due to the high linearity requirement of feedback DAC [7]. The non-linearity in a multi bit feedback DAC increases the noise floor in the signal output spectrum of the modulator and reduces the dynamic range. This noise added is due to the mismatch error introduced by the uneven spacing of DAC levels. Certain noise shaping techniques [8] have been used to reduce this non-linearity. These techniques make use of noise filtering mechanism in the signal band and move the noise power to out-of-band frequencies [9]. Several Element matching techniques like [10] [11] have been proposed to reduce the DAC non-linearity. But, they suffer from aliasing of DAC distortion errors in to signal band [7]. Though other techniques like Individual Level averaging (ILA) [12] and bi-DWA [13] are also available, but results in increased noise floor in the signal band of the output spectrum. The dynamic element matching technique which addresses the problems suffered by the above techniques is DWA [7]. The modulator system has been designed by systematic design centering method [14] for obtaining the loop-filter coefficients, by including the effect of the integrator non-idealities.

This paper presents the design and implementation of a 3rd order CT-$\Delta\Sigma$ ADC with a 4-bit quantizer targeted for audio applications. To correct the non-linearity of the DAC in the system a 4-bit DWA has been implemented using the standard cell based layout to optimize power. To achieve low power implementation loop filter architecture having minimum power consumption has been chosen.

The paper is organized as follows. We present the CT-$\Delta\Sigma$ modulator in Section II. The circuit design is described in Section III. The Simulation result along with fabricated chip is presented in Section IV, followed by the conclusion.

II. MODULATOR SYSTEM DESIGN

Figure 2 shows the CT-ΔΣ modulator with cascade of resonator in feed-forward (CRFF) [3] summation to implement third order loop filter with a multi-bit quantizer. The main advantage of feed-forward architecture is that the cascaded integrators ($I_1(s)$, $I_2(s)$, $I_3(s)$) handle only a part of the input signal. And the integrators can have a relaxed dynamic range and scaling requirements. The poles of the loop filter, H(s) are the zeros of the NTF are placed at dc. The NTF high pass response as shown in figure 3. And by adding the local feedback loops (g) around pair of integrators, NTF zeros are moved away from dc and spread over the signal band. As a result both the in band noise and dynamic range of the modulator is improved [15].

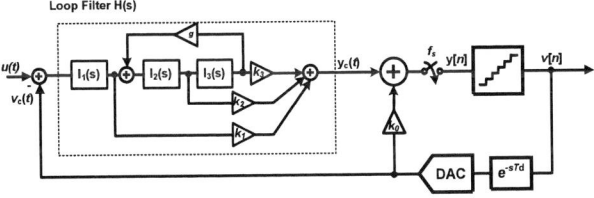

Figure 2 Block diagram of the CT-ΔΣ modulator employing a third-order

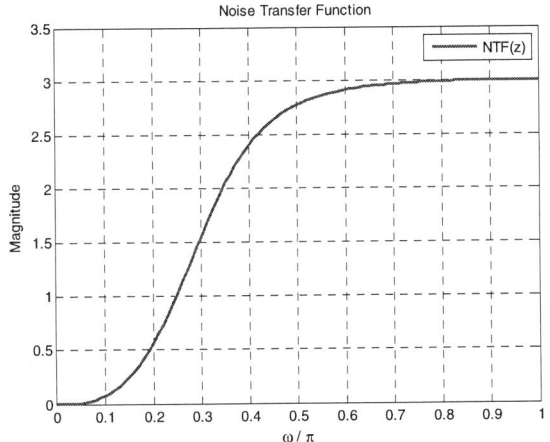

Figure 3 Noise Transfer function.

In CT-ΔΣ modulator, more than 70% of the power is consumed by the operational amplifier (op-amp), which is a key design block in the loop-filter. In order to ensure low power operation of the modulator, bias currents in the op-amp need to be minimized for the given specification. Any decrease in the op-amp bias currents decreases the unity-gain frequency f_{un}, which in turn adds excess loop-delay in the integrator. Especially, in a multiple-feedback design, it is necessary to make sure that the extra pole in the op-amp does not affect the normal operation of CT-ΔΣ modulator. In addition, the finite DC gain and fun of the op-amp impairs the characteristics of the loop-filter and thus modifies the resulting NTF [16], [17]. In this paper, a systematic design centering method explained in [16], [17] has been used for obtaining the loop filter coefficients by including the effect of the integrator non-idealities.

III. CIRCUIT DESIGN

In this section, a detailed description of circuit level blocks used in the CT- ΔΣ modulator is described.

A. Loop Filter

A third-order loop filter has been implemented with an oversampling ratio (OSR) of 64 and an achievable signal bandwidth of 48 kHz sampled at 6.144 MHz with an OBG of 3. A complete loop-filter schematic is shown in figure 4.

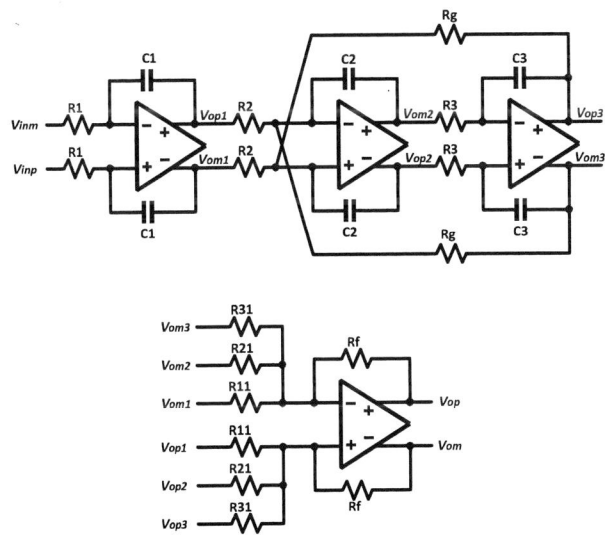

Figure 4 Loop Filter of the designed third-order CT- ΔΣ modulator.

The feed-forward co-efficient's (k_0, k_1, k_2, k_3) and summation are implemented using R_{11}, R_{21}, R_{31}, R_f resistors and a summing amplifier respectively. The input-referred noise of the loop filter is significantly dominated by the thermal noise from the input resistors and input differential pair of the first op-amp [3]. In order to mitigate the thermal noise effect from input resistor (R_1), the large value (80 KΩ) has been chosen. Also, an appropriate g_m is chosen to mitigate the thermal noise contribution from input transistors. Further, to achieve a low1/f noise, PMOS input stage is used in first integrator of an op-amp whose design is discussed in section III-B [16].

B. Operational Amplifier

Figure 5 shows the schematic of the miller compensated two-stage op-amp. A PMOS differential pair is used as the first stage and class-AB buffer is used as the second stage. M1 to M6 are long length devices to mitigate the input referred flicker noise [16]. Here, Cc and Rc are the miller compensation capacitor and zero-nulling resistor whose values are chosen as 200 fF and 10 KΩ respectively. The first stage is biased to draw 10 uA from the supply. To ensure the op-amp common mode output voltage is held at VCM, a common mode feedback (CMFB) loop is used in both stages of the op-amp (See figs. 6(a), (b)). The total current drawn by the first operational amplifier including the CMFB circuitry is 556 uA. The output of first stage (V_{o1p} and V_{o1m}) in figure 5 is fed to the input transistors M7 and M10 in figure 6(a), which averages and compares the resultant voltage with the Vref to tune the voltage (V_{CMFB1}). The CMFB circuit shown in the figure 6(b) adds current to keep the output node (V_{outp} and V_{outm}) at V_{CM}.

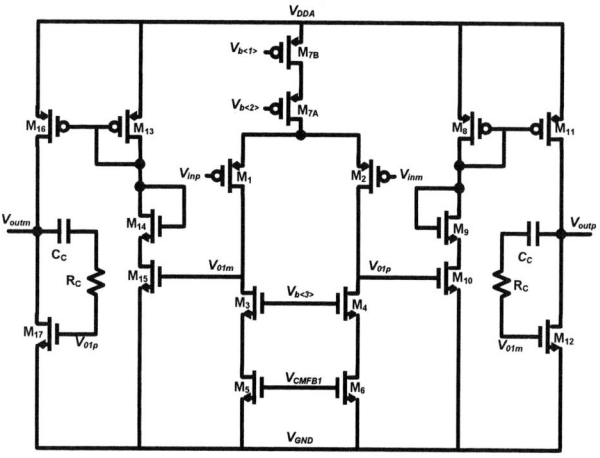

Figure 5 Operational amplifier used in integrator.

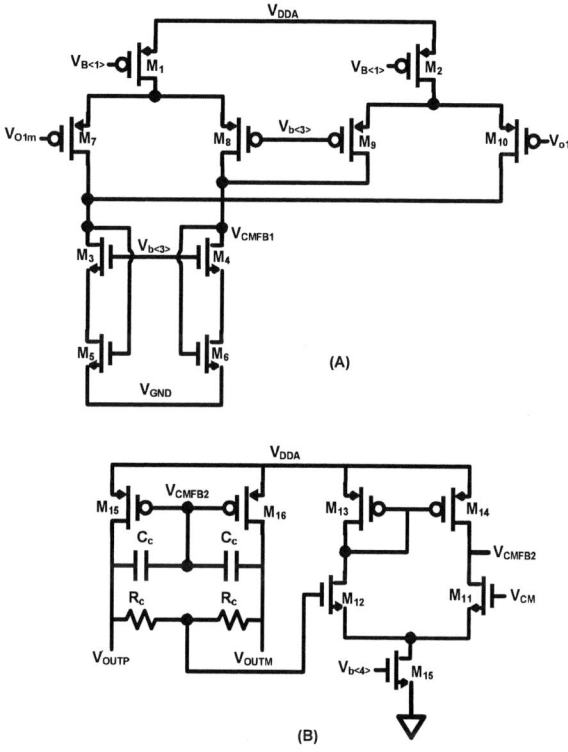

Figure 6 (A) CMFB circuit for the first stage and (B) second stage CMFB.

C. Comparator

The comparator used in the quantizer is shown in the figure 7. The comparator [18] is designed to provide sufficient regenerative gain and to avoid the effects of metastability. Here, the first stage uses a differential amplifier for reference subtraction. The amplifier is loaded with cross-coupled PMOS latches to provide initial regeneration followed by a clocked latch.

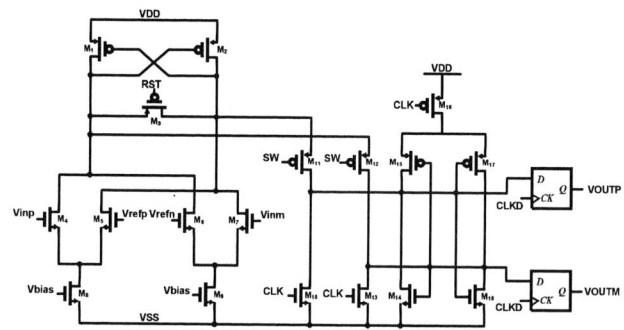

Figure 7 Comparator Circuit.

The second stage latch provides a large regenerative gain and also resolves the outputs to full logic level. The latch is disconnected from the input to avoid kickback noise. A reset phase is used to remove memory in the latches. The comparator has a resolution of 1 uV and a delay of 2 ns. The comparator is biased with a current of 26 uA and approximately dissipates 2.080 mW power at a clock rate of 6.144 MHz.

D. DAC and DWA

Though different architectural implementations of DAC are available in the literature viz., resistive DAC and switched-capacitor DAC. However, a current steering DAC [19] shown in figure 8 greatly reduces the slew requirement of first op-amp in the loop-filter which improves the stability of the system [17] has been designed for a current of 1 uA. The DWA has been implemented by a 4 layer barrel shifter method [20].

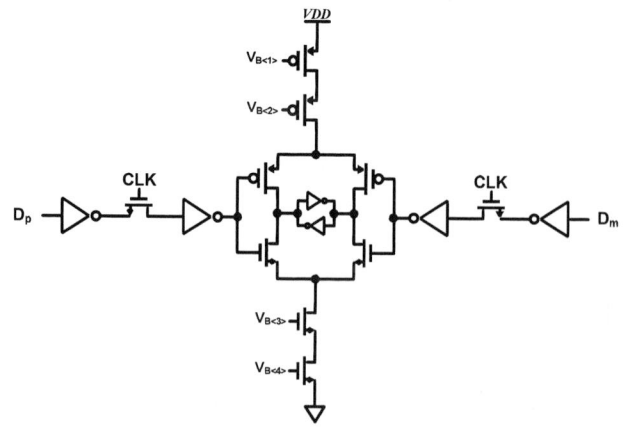

Figure 8 Current steering DAC.

978-1-4577-1735-2/12 $31.00 © 2012 IEEE

IV. SIMULATION RESULTS

The third-order CT-$\Delta\Sigma$ ADC has been designed and taped out for fabrication in the 0.5um C5FN-AMI CMOS process. Figure 9 shows the fabricated chip of the design with 4.5 mm^2. The measured SNDR is using a 2.3 kHz input tone is 85 dB. Figure 10 shows the simulated output spectrum. Table I summarizes the performance of the modulator.

TABLE I.

MODULATOR PERFORMANCE SUMMARY

Process	0.5um
Supply voltage	5 V
Sampling rate	6.144 MHz
OSR	64
Quantizer resolution	4 bit
Power dissipation	5.44 mW
SNR $_{max}$	85 dB
FoM	3.73 pJ/conv

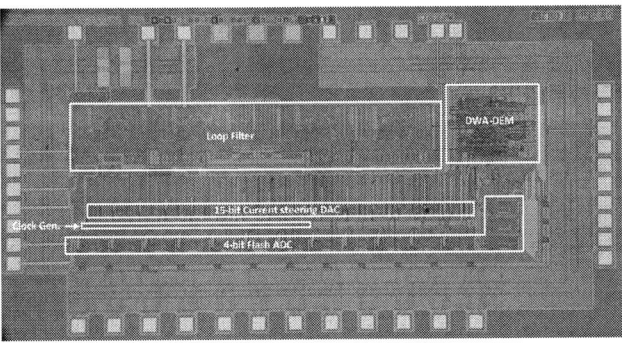

Figure 9 CT- $\Delta\Sigma$ fabricated chip picture.

V. CONCLUSION

A low-power, CT-$\Delta\Sigma$ modulator was designed in a 0.5um C5FN AMI CMOS process for data conversion employed in audio applications. The CT loop-filter coefficients were systematically obtained by incorporating the op-amp non-idealities. This method resulted in robust modulator NTF and lower static currents in the op-amps. The simulation results of the CT-$\Delta\Sigma$ exhibit a peak SNDR of 85 dB, with a maximum stable input (MSA) of -2.14 dBFS. The modulator dissipates around 5.44 mW power from a 5 V supply and achieves a FoM of 3.731 pJ/bit.

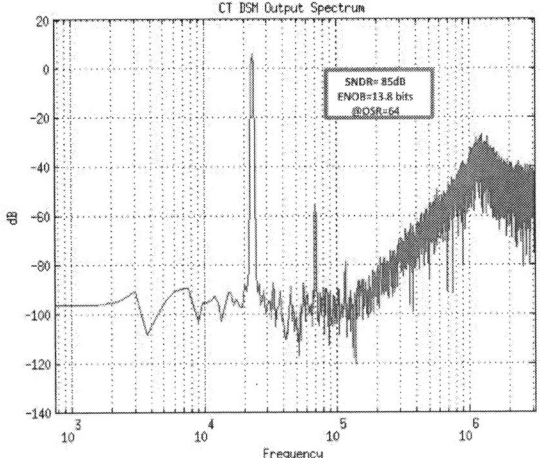

Figure 10 Simulated output spectrum.

REFERENCES

[1] S. Pavan, N. Krishnapura, R. Pandarinathan, and P. Sankar, "A power optimized continuous-time delta-sigma ADC for audio applications," IEEE Journal of solid-State Circuits, vol. 43, pp. 351–360, 2008.

[2] J. Candy, "A use of double integration in sigma delta modulation," IEEE Transcation on communcications, vol. 33, pp. 249–258, 1985.

[3] R. Schreier and G. Temes, Understanding Delta-Sigma Data Converters.IEEE press Piscataway, NJ, 2005.

[4] J. Candy and G. Temes, Oversampling Delta-Sigma Data Converters(Theory, Design, and Simulation), 1991.

[5] L. Breems, R. Rutten, and G. Wetzker, "A cascaded continuous-time delta-sigma modulator with 67-dB dynamic range in 10-MHz bandwidth," IEEE Journal of Solid-State Circuits, vol. 39.

[6] J. Kenney, "Design of multibit noise-shaping data converter," Analog integrated circuits and signal Processing, vol. 3, pp. 259–272, 1993.

[7] T. Rex and S. Terri, "Linearity Enhancement of Multibit delta-sigma A/D and D/A converters using Data Weighted averaging," IEEE Transcation on circuits and system–II, vol. 42, pp. 753–762, 1995.

[8] Shui, R. Schreier, and F. Hudson, "Mismatch shaping for a currentmode multibit delta-sigma DAC," IEEE Journal of solid-State Circuits, vol. 34, pp. 331–338, 1999.

[9] S. Norsworthy, R. Schreier, and G. Temes, Delta-Sigma Data Converters Theory, Design, and Simulation. IEEE press Piscataway, NJ, 1997.

[10] B. Leung and S. Sutarja, "Multibit delta-sigma converter incorporating a novel class of dynamic element matching techniques," IEEE Transcation on circuits and system–II, vol. 39, pp. 35–51, 1992.

[11] L. Carley, "A noise shaping coder topology for 15+ bit converters," IEEE Journal of Solid-State Circuits, vol. 24, pp. 267–273, 1989.

[12] F. Chen and B. Leung, "A high resolution multibit sigma-delta modulator with individual level averaging," IEEE Journal of Solid-State Circuits, vol. 30.

[13] I. Fujimori, L. Longo, A. Hairapetian, K. Seiyama, S. Kosic, J. Cao,and C. Shu-Lap, "A 90-dB SNR 2.5-MHz output-rate ADC using cascaded multibit delta-sigma modulation at 8x oversampling ratio," IEEE Journal of Solid-State Circuits, vol. 35.

[14] S. Pavan and P. Sankar, "Power Reduction in Continuous-Time Delta-Sigma Modulators Using the Assisted Opamp Technique," IEEE Journal of solid-State Circuits, vol. 45, pp. 1365–1379, 2010.

[15] M. Ortman and F.Gefers, Continuous-time Sigma-Delta A/D conversion; fundamentals, performance limits and robust implementations. springer, 2006.

[16] S. Balagopal, R. Roy, and V. Saxena, "A 110uW single-bit continuous time delta-sigma converter with 92.5dB dynamic range," IEEE Dallas circuits and systems workshop (DCAS).

[17] K. Reddy and S. Pavan, "A 20.7 mW continuous-time $\Delta\Sigma$ modulator with 15MHz bandwidth and 70 dB dynamic range," in Solid-State Circuits Conference, 2008. ESSCIRC 2008. 34th European. IEEE, 2008, pp. 210–213.

[18] G. Mitteregger, C. Ebner, S. Mechnig, T. Blon, C. Holuigue, and E. Romani, "A 20-mW 640-MHz CMOS Continuous-Time Sigma-Delta ADC With 20-MHz Signal Bandwidth, 80-dB Dynamic Range and 12-bit ENOB," Solid-State Circuits, IEEE Journal of, vol. 41, no. 12, pp.2641–2649, 2006.

[19] K. Reddy and S. Pavan, "A power efficient continuous time Sigma-Delta modulator with 15 MHz bandwidth and 70 dB dynamic range," Analog Integrated Circuits and Signal Processing, vol. 63, pp. 397–406, 2010.

[20] S. Paton, A. Di Giandomenico, L. Hernandez, A. Wiesbauer, T. Potscher, and M. Clara, "A 70-mW 300-MHz CMOS continuous-time sigma-delt a ADC with 15-MHz bandwidth and 11 bits of resolution," Solid-State Circuits, IEEE Journal of, vol. 39, no. 7, pp. 1056–1063, 2004.

Two Techniques to Reduce Gain and Offset Errors in CMOS Image Sensors using Delta-Sigma Modulation

Kuangming Yap and R. Jacob Baker

Department of Electrical and Computer Engineering
Boise State University
Boise, ID, U.S.A.

Abstract—**A per-column, delta-sigma, analog-to-digital converter for use in CMOS image sensors is reported. Two techniques, subtraction and preconditioning, are proposed to compensate for the column-to-column mismatches and the resulting fixed-pattern noise introduced into the image. Equations governing the operation of the proposed topology are developed. Experimental results verify that even in the presence of very large offsets, such as a 200 mV mismatch in the MOSFETs' threshold voltages, the proposed topology operates as desired.**

Keywords-Delta-Sigma Modulator; DSM; CMOS Image Sensor ADC; Gain Error

I. INTRODUCTION

This paper presents a per-column analog-to-digital converter (ADC) design using delta-sigma modulation (DSM) for use in a CMOS imager. Currently, mass-produced CMOS imagers use ADCs that are either pipeline or column parallel architectures requiring precision components [1]. The benefit of using a DSM ADC is that it's more tolerant to noise and power supply variations than traditional analog-to-digital conversion techniques. Also DSM-based ADCs don't require precise components, so they can be manufactured with good yield. However, as with any per-column ADC, there are practical matching concerns [1], [2], [3]. For example, fixed pattern noise will occur in the decoded image if the transfer functions of each column-parallel ADC are not identical. Variations in device characteristics can cause gain and offset errors in the ADCs' transfer functions. Path switching techniques have been introduced to reduce the effects of these errors [3], [6]. Unfortunately, a threshold voltage mismatch in the reference source-follower transistor in these topologies will cause a gain error in the DSM ADC transfer function. This gain error can be reduced through two different techniques, subtraction and preconditioning. Both techniques are discussed in this paper.

II. GAIN ERROR CORRECTION BY MEANS OF SUBTRACTION

Threshold voltage mismatches result in gain errors that can be reduced by generating two different reference currents through the same reference source-follower transistor, M14 in Fig. 1 [5], [6]. One reference current, say I_{REF1} in Fig. 1, is directed to one side of the DSM ADC, while the other reference current, I_{REF2}, is directed to the other side. The current mirror, M7/M8, subtracts out the reference source-follower transistor threshold voltage component resulting in a reference current that is free from the influence of M14's

threshold voltage. This method is known as gain error correction by means of subtraction, and it's shown schematically in Fig. 1.

This DSM ADC measures the difference between the two analog input signals, V_{RESET} and V_{IMAGE}, with respect to two reference input signals, V_{REF1} and V_{REF2}, and converts this difference to a train of digital pulses. The conversion period takes N clock cycles. This DSM ADC requires a 4-phase, non-overlapping clock signal. The PHI1, PHI2, PHI3, and PHI4 signals are the four non-overlapping clock phases and their complement signals are PHI1B, PHI2B, PHI3B, and PHI4B, respectively. These signal's frequency is 1/4 the rate of the master clock denoted as f_{PHI}.

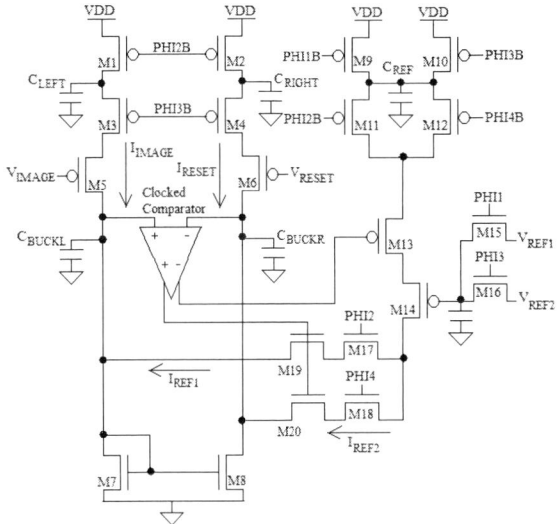

Figure 1. DSM ADC with gain error correction by means of subtraction [5], [6].

M5 and M6 are respectively known as the image and reset source-follower transistors. A dummy capacitor is added to the gate of M14 as a means to reduce the effects of charge injection and clock feed-through when PHI1 and PHI3 signals transition from high to low. The reference signal voltage needs to be on the gate of M14 a phase earlier and stays unchanged for the whole duration of the subsequent phase. This is to prevent any error in the magnitude of the two reference currents, I_{REF1} and I_{REF2}.

978-1-4577-1735-2/12 $31.00 © 2012 IEEE

On the first clock phase, M9 turns on and C_{REF} is charged to VDD. At the same time, M15 turns on and allows the first reference voltage, V_{REF1} to propagate to the gate of M14. At the end of the first clock phase, the clocked comparator measures the voltages on capacitor C_{BUCKL} and C_{BUCKR} and turns on M13, M19 and M20 for the remaining three clock phases if the voltage on C_{BUCKR} is lower than the voltage on C_{BUCKL}. M13, M19 and M20 are turned on for M times over the entire conversion period. During the second clock phase, M11 turns on and the first reference current, I_{REF1}, flows from C_{REF} to C_{BUCKL} if M13, M19 and M20 are turned on. The average I_{REF1} that flows into C_{BUCKL} is

$$I_{REF1} = \frac{M}{N} C_{REF} \frac{f_{PHI}}{4}(VDD - V_{REF1} - V_{th,M14}). \quad (1)$$

On the same clock phase, the voltages on C_{LEFT} and C_{RIGHT} are set to VDD. Then the third clock phase, C_{REF}, is reset back to VDD and the second reference voltage signal, V_{REF2}, is propagated to the gate of M14. During this phase, the image current, I_{IMAGE}, and reset current, I_{RESET}, will flow to C_{BUCKL} and C_{BUCKR} respectively.

$$I_{IMAGE} = C_{LEFT} \frac{f_{PHI}}{4}(VDD - V_{IMAGE} - V_{th,M5}) \quad (2)$$

$$I_{RESET} = C_{RIGHT} \frac{f_{PHI}}{4}(VDD - V_{RESET} - V_{th,M6}) \quad (3)$$

During the last clock phase, the second reference current, I_{REF2}, will flow from C_{REF} to C_{BUCKR} if M13, M19, and M20 remained on. The average I_{REF2} that flows into C_{BUCKR} is

$$I_{REF2} = \frac{M}{N} C_{REF} \frac{f_{PHI}}{4}(VDD - V_{REF2} - V_{th,M14}). \quad (4)$$

I_{REF1} has the same M14 voltage threshold component in its current equation as I_{REF2}. Assuming that there is no threshold voltage offset between the current mirror transistors, M7 and M8, this current mirror will subtract out M14's threshold voltage component from the true reference current, and then its average magnitude is

$$I_{REF} = I_{REF2} - I_{REF1} \quad (5)$$

$$I_{REF} = \frac{M}{N} C_{REF} \frac{f_{PHI}}{4}(V_{REF1} - V_{REF2}) \quad (6)$$

noting that mismatches in M7 and M8 are corrected for in the same way as mismatches in M5 and M6.

The digital code representation of the analog input signals, with respect to the two reference signals, can be found by summing the current into the C_{BUCKR} capacitor.

$$M = N \left(\frac{C_{LEFT}(VDD - V_{IMAGE} - V_{th,M5}) - C_{RIGHT}(VDD - V_{RESET} - V_{th,M6})}{C_{REF}(V_{REF1} - V_{REF2})} \right) \quad (7)$$

The ADC is tolerant to gain errors caused by threshold voltage mismatches because the denominator of the transfer function is free from the influence of the threshold voltage. Note that the gain of the DSM ADC can be changed by controlling the difference between the two reference voltage signals, V_{REF1} and V_{REF2}.

This DSM ADC is still susceptible, however, to a path offset error. The ADC can be modified to accommodate a path switching technique to reduce the effects of offset errors in its transfer function. Fig. 2 illustrates the DSM ADC with gain

error correction by means of subtraction and offset error correction.

The conversion period of the DSM is now divided into 2 equal halves, where each half is $N/2$ clock cycles long. On the first half of the sensing period, control signals SLT and SLB are set to VDD and ground, respectively. During this sensing period, the digital code representation of the analog input signals, with respect to the two reference input signals, $M_{t=1}$, is

$$M_{t=1} = \frac{N}{2} \left(\frac{C_{LEFT}(VDD - V_{IMAGE} - V_{th,M5}) - C_{RIGHT}(VDD - V_{RESET} - V_{th,M6})}{C_{REF}(V_{REF1} - V_{REF2})} \right). \quad (8)$$

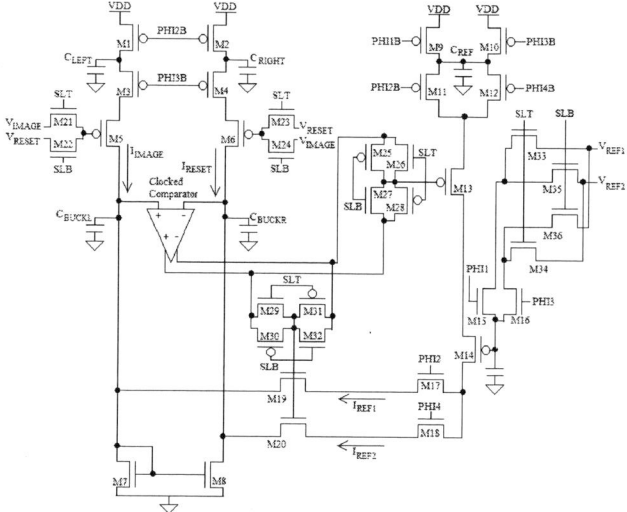

Figure 2. DSM ADC with both gain and offset error correction. Gain error correction by means of subtraction.

During the next half of the conversion period, SLT and SLB are driven to ground and VDD, respectively. The digital code representation of the analog input signals, with respect to the two reference input signals, for the second half of the sensing period, $M_{t=2}$ is

$$M_{t=2} = \frac{N}{2} \left(\frac{C_{RIGHT}(VDD - V_{IMAGE} - V_{th,M6}) - C_{LEFT}(VDD - V_{RESET} - V_{th,M5})}{C_{REF}(V_{REF1} - V_{REF2})} \right). \quad (9)$$

At the end of the sensing period, the digital output code for the two halves of the sensing period are added together and the final digital output code for the DSM with both gain and offset correction, is

$$M_{t=1} + M_{t=2} = \frac{N}{2} \left(\frac{(C_{LEFT} + C_{RIGHT})(V_{RESET} - V_{IMAGE})}{C_{REF}(V_{REF1} - V_{REF2})} \right). \quad (10)$$

The input-output transfer function of this DSM does not contain the threshold voltage of any transistor in the DSM. This means that the DSM is robust to any offset or gain error caused by threshold voltage mismatches. However, a gain error will still occur if the capacitors C_{LEFT}, C_{RIGHT}, and C_{REF} are mismatched. Note that the least significant bit voltage, V_{LSB} for this DSM ADC, with both offset and gain error correction by means of subtraction, is

$$V_{LSB} = \frac{2}{N} \left(\frac{C_{REF}}{(C_{LEFT} + C_{RIGHT})} \right)(V_{REF1} - V_{REF2}). \quad (11)$$

The bit accuracy increases linearly with the number of clock cycles, N, during the sensing period.

III. DELTA-SIGMA MODULATION ADC WITH GAIN ERROR CORRECTION BY MEANS OF PRECONDITIONING

At the beginning of every clock period, the initial voltage on the reference capacitor is usually set to VDD. If the initial voltage on C_{REF} is preconditioned by the reference source-follower transistor to another reference voltage instead of VDD, the true reference current that flows into C_{BUCKR} will be free from the influence of the threshold voltage. This will then lead to a transfer function that is robust to gain error. Fig. 3 shows the DSM ADC with gain error correction by means of preconditioning.

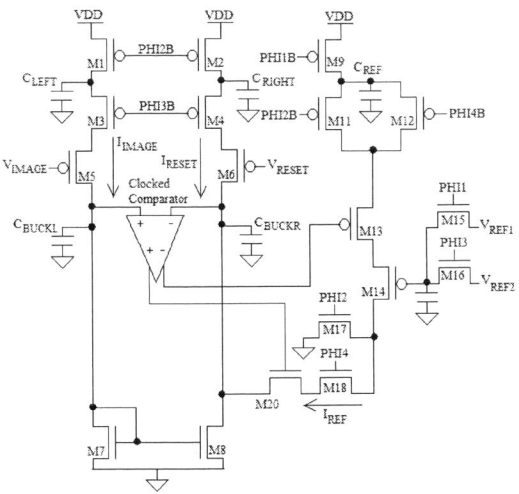

Figure 3. DSM ADC with gain error correction by means of preconditioning.

The operation of this DSM ADC is very similar to the DSM ADC with gain error correction by means of subtraction. It also requires 4-phase non-overlapping clock signals and its clock frequency is also 1/4 the rate of the master clock denoted as f_{PHI}. The generation of the image current, I_{IMAGE}, and reset current, I_{RESET}, in this DSM ADC is similar. However, the generation of the reference current is different. On the first phase of the clock signal, C_{REF} is set to VDD and the first reference voltage, V_{REF1}, propagates to the gate of M14. At the end of the first clock phase, the clocked comparator turns on M13 and M20 for the remaining clock phases, if the voltage on C_{BUCKR} is lower than the voltage on C_{BUCKL}. Like before, this happens M times over the entire conversion period. On the next clock phase, the voltage on C_{REF} is set to $V_{REF1} + V_{th,M14}$, if M13 is turned on by the clocked comparator. This is accomplished by allowing a current to flow from C_{REF} to ground through M11, M13, M14, and M17. The second reference current, V_{REF2}, will then propagate to the gate of M14 on the following clock phase. On the last clock phase, the reference current, I_{REF}, will flow from C_{REF} to C_{BUCKR}, if M13 and M20 are turned on by the clocked comparator. The magnitude of I_{REF} is

$$I_{REF} = \frac{M}{N} C_{REF} \frac{f_{PHI}}{4} \left(V_{REF1} + V_{th,M14} - V_{REF2} - V_{th,M14} \right) \quad (12)$$

which simplifies to

$$I_{REF} = \frac{M}{N} C_{REF} \frac{f_{PHI}}{4} \left(V_{REF1} - V_{REF2} \right). \quad (13)$$

The I_{REF} equation for this topology is similar to the I_{REF} equation of the topology with gain error correction by means of subtraction. Therefore, the same gain error correction property and input-output transfer function can be used. As before, a modification can be made to incorporate the path switching technique to reduce the effects of offset error in the transfer function. Its schematic for this design is shown in Fig. 4.

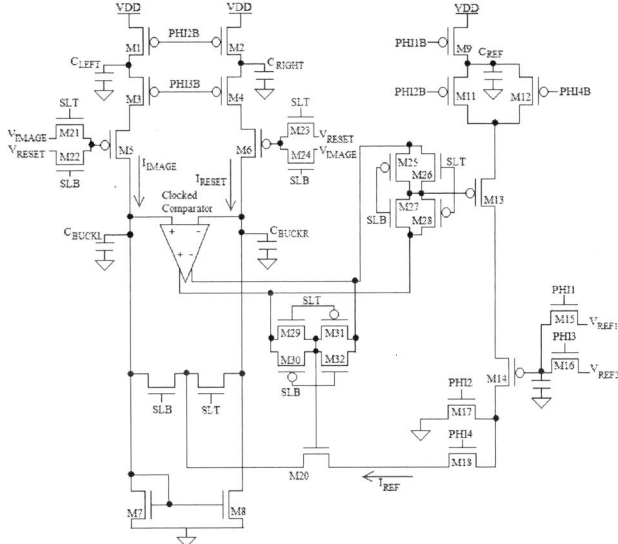

Figure 4. DSM ADC with both gain and offset error correction. Gain error correction by means of preconditioning.

IV. TEST CHIP RESULTS

A test chip was fabricated in On's C5 process containing test structures to evaluate the performance of the proposed ADCs (see Fig. 5). The chip contains column slices with inputs to introduce mismatches to verify that the equations derived in this paper are valid.

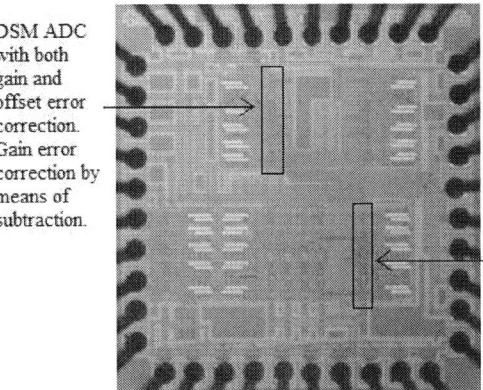

Figure 5. Test chip microphotograph.

Fig. 6 shows the measured results with offset and gain errors corrected by means of subtraction. Fig. 7 shows the test results for the DSM ADC with offset and gain errors corrected by means of preconditioning. Threshold voltage mismatch via M14 was evaluated (made worse) by offsetting V_{REF1} and V_{REF2} by equal amounts. Note that the offsets are unrealistically high in both figures, that is, 200 mV to verify robust operation.

The slope of the transfer function, using offset and gain error correction by means of preconditioning, Fig. 7, was constant with varying offsets on M14. However, using offset and gain error correction by means of subtraction, Fig. 6, had a steeper slope in its transfer function when the reference voltage is offset by −200 mV.

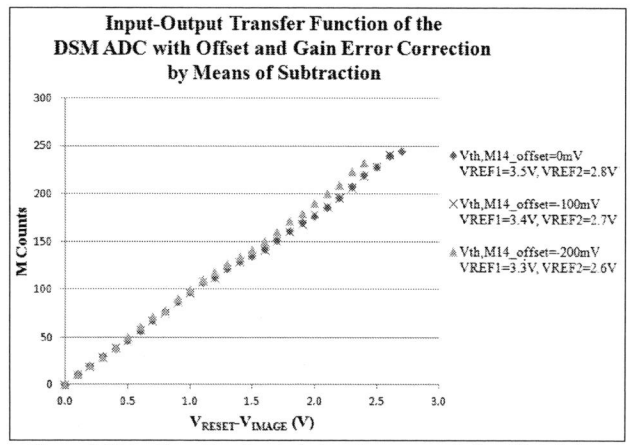

Figure 6. Using the subtraction technique to compensate for offsets in the reference voltages and threshold voltage of M14.

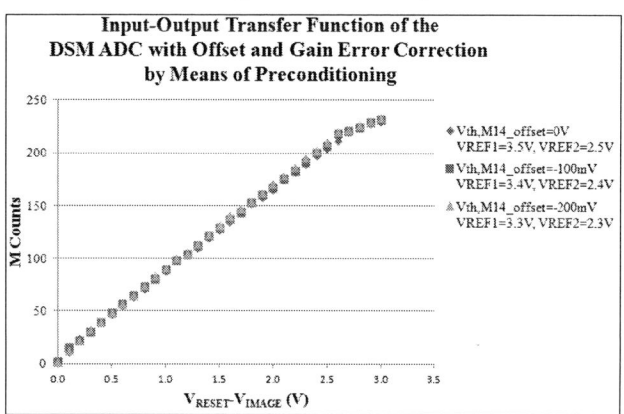

Figure 7. Using the preconditioning technique to compensate for offsets in the reference voltages and threshold voltage of M14.

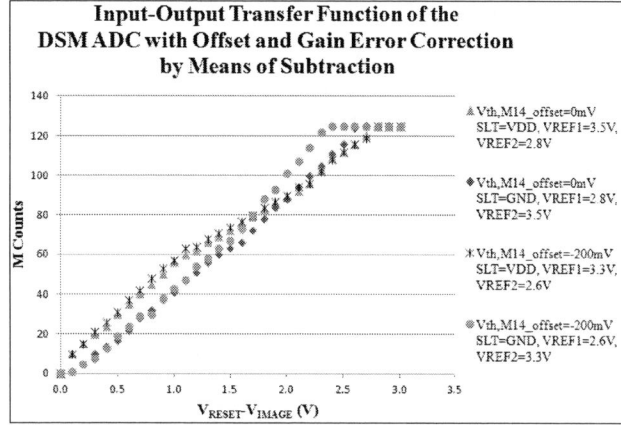

Figure 8. Showing transfer function with offset and gain errors by means of subtraction for each half of the conversion period.

Fig. 8 plots the input-output transfer function for each half of the conversion period. On the first half of the conversion period when SLT was set to VDD, the slope remained constant for both voltage offsets, 0mV and -200mV. However, a slope increase occurs during the second half of the conversion period, when SLT was set to GND and the reference voltage offset is −200 mV. This slope increase is most likely caused by the incomplete current mirroring of the larger second reference current from the left branch to the right branch of the DSM ADC. In other words, the drain of M7 must be greater than the NMOS threshold voltage, while the drain of M8 can swing near ground. This incomplete current mirroring will result in an equivalent V_{REF2} that is larger than intended, and thus resulting in a positive gain error. It should be noted, again, that a 200 mV offset is unrealistically large, so this issue, the increase in the transfer function's slope during the second half of the conversion, will likely never occur.

V. SUMMARY AND CONCLUSIONS

Gain error correction by the method of preconditioning is robust to a wide range of offsets. However, gain error correction by method of subtraction is only effective if the reference voltage levels are not too low. Path switching techniques can be applied in both methods to remove both offset and gain errors caused by mismatches in MOSFETs. Experimental results show that the matching between devices can be unrealistically large and the column parallel delta-sigma ADCs' behaviors track and match with good linearity. Finally, sensing resolution can be improved by sensing for longer periods of time. If, for example, a 100 MHz clock is used then the row readout time, using the data in Figs. 6-8 (250 counts maximum), is only 2.5 µs. For a 1,000 pixel row, this corresponds to a readout time of 2.5 ns/pixel which is faster than usually required.

ACKNOWLEDGMENT

We would like to thank the developers of the Electric VLSI design program and the MOSIS program. The Electric VLSI program was used as the design tool and the MOSIS program was used to fabricate the test chip.

REFERENCES

[1] M. F. Snoeij, A. J. P. Theuwissen, K. A. A. Makinwa, J. H. Huijsing, "Multiple-Ramp Column Parallel ADC Architectures for CMOS Image Sensors," *IEEE Journal of Solid-State Circuits*, vol. 42, pp. 2968-2977, December 2007.

[2] A. J. P. Theuwissen, "Better Pictures Through Physics," *IEEE Solid-State Circuits Magazine*, vol. 2, pp. 22-28, Spring 2010.

[3] D. Montierth, "Using delta-sigma modulation for sensing in a CMOS imager," Master Thesis, Boise State University, October 2009.

[4] R. J. Baker, "CMOS Circuit Design, Layout and Simulation, Third Edition," *Wiley-IEEE*, 2010.

[5] K. Yap and R. J. Baker, "Gain Error Correction for CMOS Image Sensor using Delta-Sigma Modulation", *Proceedings of the IEEE/EDS Workshop on Microelectronics and Electron Devices (WMED)*, pp. 52-55, April 16, 2010.

[6] K. Yap, "Gain and Offset Error Correction for CMOS Image Sensors using Delta-Sigma Modulation," Master's Thesis, Boise State University, May 2010.

High Voltage Tolerant Stacked MOSFET in a Buck Converter Application

Stacey Page, Andrew Wajda, and Herbert Hess

Electrical and Computer Engineering
University of Idaho
Moscow, Idaho USA
stacey.page@vandals.uidaho.edu

Abstract—**A Buck converter, created with a monolithic implementation of series connected MOSFETs at high-voltage, is presented. Using a single low-voltage control signal to trigger the bottom MOSFET in the series stack, a voltage division across parasitic and inserted capacitances in the circuit is used to turn on the entire stack of devices. The governing equations for the Stacked MOSFET are presented, and reliable operation for use in a Buck converter is presented in a 0.180 um Silicon-on-Insulator (SOI) CMOS process. The prototype is shown to handle 25 mA with a rail voltage of 3.3 V and a frequency range of 1-20 Megahertz.**

Keywords-Stacked MOSFET; CMOS Buck Converter;high voltage tolerant CMOS;

I. INTRODUCTION

The stacked MOSFET is a circuit intended to imitate a high voltage transistor using entirely low voltage components, and was first presented by Hess and Baker in 2000 [1]. By stacking transistors drain to source, and using some peripheral circuitry, the high voltage can be evenly split amongst the transistors. No transistor in the stack is subjected to drain to source voltage levels that it cannot support. In theory, for a given IC process, a stack of n transistors can support n times the process voltage.

The stacked MOSFET can be implemented in three topologies: the NMOS, the PMOS and the push-pull. The push-pull implements the PMOS and NMOS sections like a CMOS inverter, with the PMOS on top of the NMOS.

The goal of this paper is to present the stacked MOSFET designed in American Semiconductor Inc. (ASI's) 180 nm, 3.3 V, SOI process along with design equations and simulated results of the device in a Buck DC-DC converter implementation. This DC-DC converter could be used to power sensors, electronics, actuators, and a variety of scientific devices.

Fig. 1 shows the topology for the two device push-pull stack in the Buck converter configuration. When used in a Buck converter topology, the bottom NMOS stack replaces the standard diode function and the PMOS stack performs the switching function. A level shifter (not shown) is required to

Air Force Research Laboratory, Space Vehicles, Protection Branch (AFRL/RVSE)

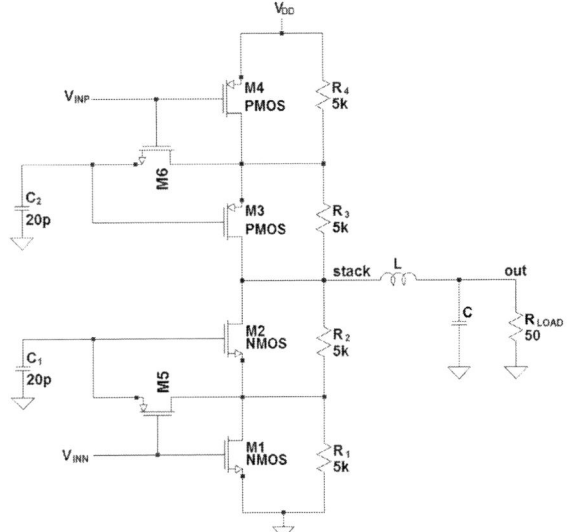

Figure 1. Schematic of push-pull Stacked MOSFET Buck DC-DC with pass gates

provide the necessary source-gate voltage, V_{INP}, to the PMOS stack to insure that it is in the saturation region when on. It should be noted that the level shifter would be subjected to the same breakdown voltages as the stack and needs to be cascoded or replaced with another device as the supply voltage increases.

II. PRINCIPLES OF OPERATION

A. General Stacked MOSFET Operation

Fig. 2 shows the topology of a two-device NMOS Stacked MOSFET. By placing MOSFETs in series and equally dividing the desired high voltage across them, for the entire switching period, reliable high voltage control can be achieved.

The topology of a Stacked MOSFET is designed to imitate a high voltage NMOS transistor, being controlled by a low-voltage logic level as demonstrated by Hess and Baker using discrete power MOSFETs. Their characterization was well suited to the discrete design process, utilizing specification

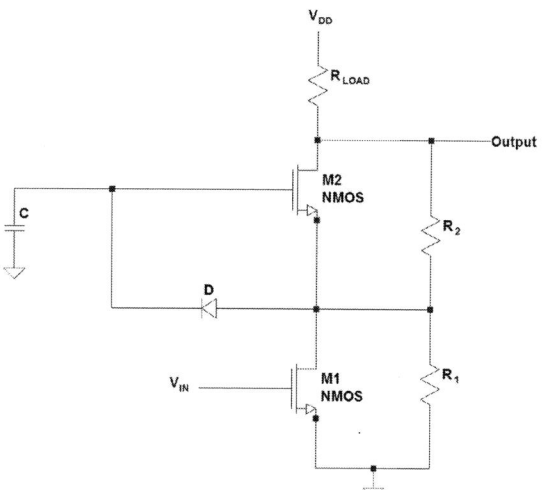

Figure 2. Schematic of NMOS Stacked MOSFET

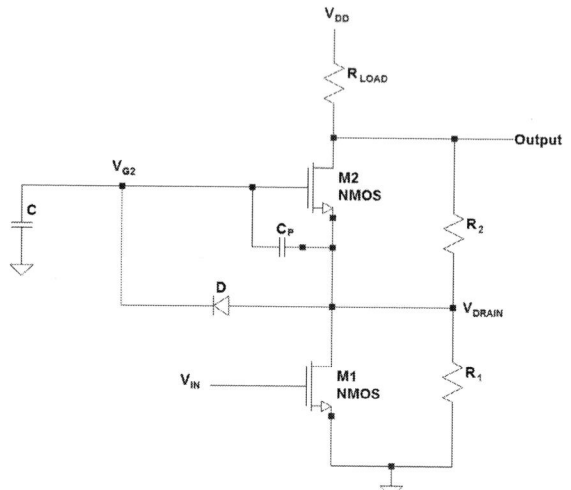

Figure 3. NMOS Stacked MOSFET, including parasitic capacitance and voltage locations

sheet parameters, such as MOSFET input and output capacitance. To realize this circuit concept in IC technology the governing equations need to be characterized for IC design parameters.

The triggering of the Stacked MOSFET is accomplished through capacitive voltage division. As shown in Fig. 3, there exists an inherent parasitic capacitance C_P between the gate and the source of M2. This capacitance, along with a capacitor C inserted in the gate leg of M2 will set the value of V_{GS2} that turns on M2.

By design, the off-resistance of each transistor is much higher than the resistance of the parallel resistors (R_1 or R_2) such that

$$R_{LOAD} << R_1 + R_2, \qquad (1)$$

and the output voltage will rise to V_{DD}. Since M1 is off, the node V_{DRAIN} is free to take on the value dictated by the voltage divider of R_1 and R_2. If R_1 and R_2 are sized equally then

$$V_{DRAIN} = V_{DD}/2. \qquad (2)$$

This voltage is greater than V_{G2}, and causes the diode to be forward biased. The resulting voltage at the gate of M2 will be

$$V_{G2} = V_{DRAIN} - V_{DIODE} = \frac{V_{DD}}{2} - V_{DIODE}, \qquad (3)$$

where V_{DIODE} is the forward voltage drop across the diode D.

In the current iteration of the stacked MOSFET the diode has been replaced by a pass gate as shown in Fig. 1. M5 and M6 are the PMOS and NMOS devices, respectively, that are used to replace the diodes. By using a transistor of the complement to the stack (NMOS in a PMOS stack and PMOS in a NMOS stack) charge is allowed to flow onto the capacitor while the stack is in the off state. When the stack turns on, the pass gate turns off trapping charge on the capacitor and gate of

the second transistor. By applying V_{IN} to the gate of the pass gate transistor we can insure that when the NMOS stack is on, the pass gate is off and vice versa. This emulates the behavior of the diode and the voltage dropped across the transistor is smaller than what an actual diode would have, which translates to a higher V_{G2}.

The off-condition, with the output voltage at V_{DD} and V_{DRAIN} at $V_{DD}/2$, exhibits static voltage balancing and results in this condition being safely held.

When V_{IN} rises high, M1 is turned on, pulling V_{DRAIN} to ground. The pass gate is off, leaving the gate-to-source voltage of M2 to be set by the capacitive voltage divider of C and C_P. C_P represents the lumped total parasitic capacitance across the gate-source of M2 and can be solved for as (4),

$$C_P = C_{DIODE} + C_{GS} + C_{GB} + C_{GD}(1 - E_{V1}) + C_{DS}(1 - E_{V2}), \qquad (4)$$

where C_{DIODE} is the capacitance of the pass gate (another parasitic MOSFET capacitance) and C_{GS}, C_{GS}, C_{GD} and C_{DS} are the corresponding MOSFET junction capacitances of M2. E_{V1} and E_{V2} are used to approximate the Miller capacitance resulting from C_{GD} and C_{DS}, and are defined in [2].

In Mentze, Hess, Buck, and Windley [2], the equations for selecting the capacitance C is derived. For a two-device Stacked MOSFET, the capacitance can be found using

$$C = C_P \left(\frac{V_{GS} + V_{DIODE}}{\frac{V_{DD}}{2} - (V_{GS} - V_{DIODE})} \right). \qquad (5)$$

B. Two Device Stacked MOSFET

Using the previous equations, a two-device Stacked MOSFET has been designed and simulated for implementation a 180 nm, SOI process. This process has a breakdown voltage of 1.8 V and a supply voltage of 3.3 V.

Each MOSFET in the stack shown in Fig. 1 is sized to have a width of 800 μm and a gate-source voltage of 1.65 V. The parasitic capacitances, under the desired operating conditions, can be extracted from the device models as shown in Table I. This table also includes the extracted diode junction (pass gate) capacitance at the appropriate biasing

Table I. CALCULATED JUNCTION CAPACITANCES

Capacitance	*Calculated Value (fF)*
Gate-Source	1297.00
Gate-Bulk	0.00
Gate-Drain	318.75
Drain-Source	0.00
Pass Gate	24.18

conditions. Accordingly, C can be sized using equations (4) and (5) to be 430.77 fF for the NMOS stack. Fig. 1 shows C_1 and C_2 to be 20 pF for the NMOS and PMOS stacks. Because the process provided no explicit capacitors, MOSFET capacitors were created. As the frequency increases the MOSFET capacitors do not exhibit strong inversion and the capacitance decreases. This occurs because the two methods of electron generation, diffusion of minority carrier electrons from the p-type substrate across the space charge region and thermal generation of electron-hole pairs within the space charge region cannot change instantaneously as described in [3]. To insure that the necessary capacitance was available to supply charge to the gate for turn on, the capacitors were oversized to compensate.

III. SIMULATION RESULTS

The simulated drain-to-source voltages resulting from the previous design values, at a frequency of 1 MHz in the push pull topology are shown in Fig. 4. The voltages are evenly distributed, causing no device to exceed the drain-to-source breakdown voltage of 1.8 V. All of the transistors in the stack are turning on and off correctly and the resistive-capacitive (RC) delay is negligible at this frequency.

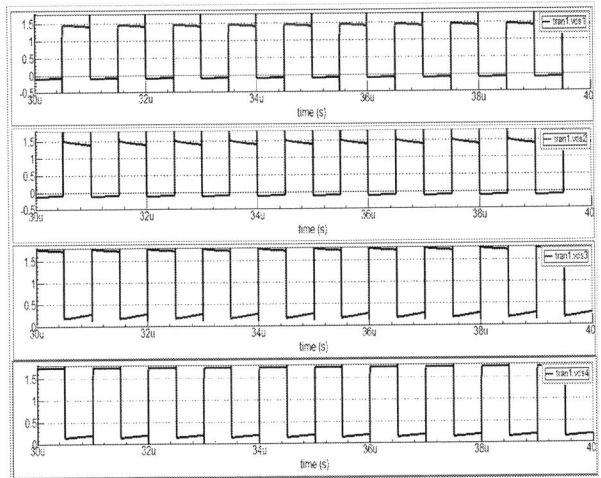

Figure 4. Drain-source voltages of push-pull Stacked MOSFET simulation-1 MHz

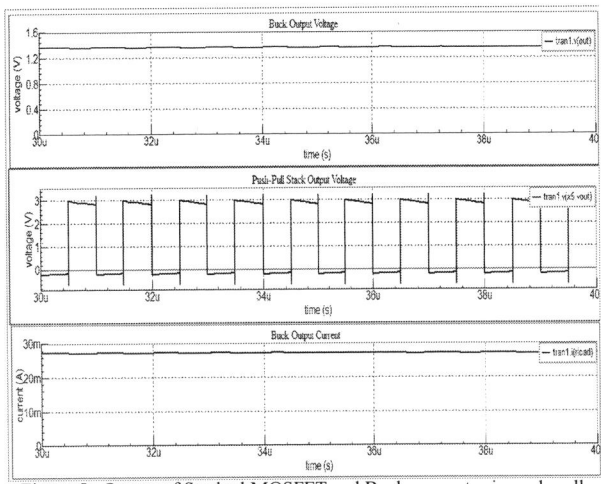

Figure 5. Output of Stacked MOSFET and Buck converter in push pull configuraion-1 MHz

In Fig. 5 we see that at 1 MHz switching frequency, inductance of 100 μH, capacitance of 100 nF, and a load resistance of 50 Ω, we get an average output voltage of around 1.36 V and a current of 27 mA with 50% duty cycle and minimum ripple. The middle waveform is showing the output voltage of the push pull-stack to the passive components which is about 3.1 V, peak to peak, and has a slight drop in voltage during the on period, because of the difficulty of keeping the pass gates turned off with the capacitors behind them.

Fig. 6 shows the drain-to-source voltages of the stack at 20 MHz switching frequency. The transistors in the stack are still turning on and off, but the RC time constants are beginning to have a significant effect. The RC time constants show up in the turn-on and turn-off of the PMOS and NMOS stacks, respectively. The interior transistors (M2 and M3) again show that there is some form of leakage from the capacitors, and it is more pronounced at the higher frequencies. They are

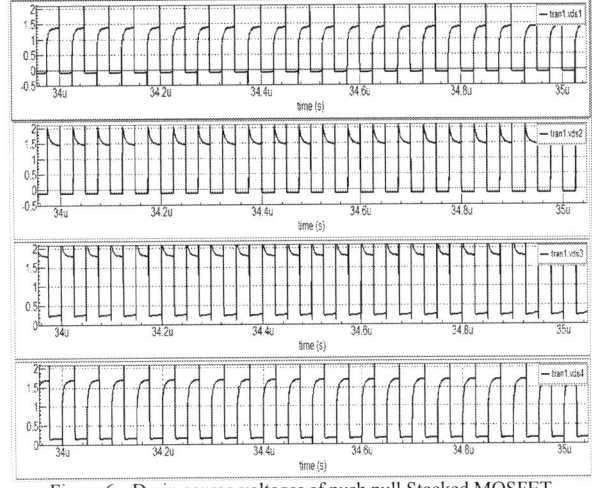

Figure 6. Drain-source voltages of push pull Stacked MOSFET simulation-20 MHz

turning on and off, but are not holding their voltage levels as well as they did at 1 MHz switching frequency. The highest voltage seen is 2.0 V and is on for about a quarter of a pulse before it drops below the breakdown voltage of 1.8 V.

The output voltage from the stack is reduced slightly at higher frequencies, from 3.3 V to 3.0 V and the slope during the on-period is much more pronounced. At 20 MHz we are approaching the limit of where the device can function while providing the desired output through the Buck converter. This shows that the push-pull stack is still statically and dynamically balanced during the entire switching period. The push-pull stack has active rising and falling edges, because of the active pull up of the PMOS stack and the active pull down of the NMOS stack.

Fig. 7 shows the output current from the Buck converter with an inductance of 10 µH, a capacitance of 10 nF, and a load resistance of 50 Ω. The average current value (bottom waveform) is 26 mA with an output voltage from the stack (center) of 3.0 V with some loss during the on time.

The output voltage of the buck converter (top) has minimal ripple voltage and holds steady at about 1.31 V. This is a little less than the 1.65 V expected for a 50% duty cycle but is still in

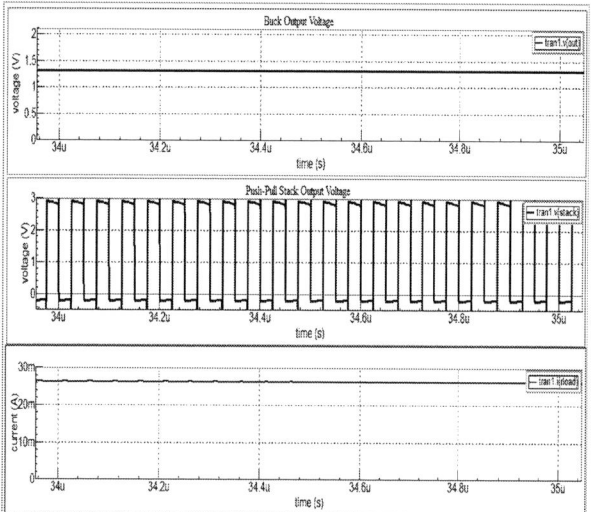

Figure 7. Output of Stacked MOSFET and Buck converter in push pull configuraion-20 MHz

the acceptable range when compared to previous versions developed in [5]. Some increase could be applied to the duty cycle to reach half of the supply voltage.

IV. CONCLUSION

This paper presented the design equations and simulation results for a two-device stack in NMOS, PMOS, and push-pull topologies in a 180 nm SOI process. The simulations showed that the stacked MOSFET can handle high-voltage switching in a Buck DC-DC converter application.

All of the stacked MOSFET topologies showed monolithic switching capabilities in simulations and experiments with static and dynamic balancing of the output signal. The stacked MOSFET has been realized in discrete and integrated configurations in several different processes and simulations, and DC-DC converters could be used to power sensors, electronics, actuators and a variety of scientific devices.

REFERENCES

[1] H. Hess and R.J. Baker, "Transformerless capacitive coupling of gate signals for series operation of power MOS devices," IEEE Trans. Power Electron, vol. 15 no. 5, pp. 923-930, Sep. 2000.

[2] Mentze, E, Hess, H., Buck, K, Windley, T, "A Scalable High-Voltage Output Driver for Low-Voltage CMOS Technologies," IEEE Trans. Very Large Scale Integ.(VLSI) Systems, vol. 14, no. 12, pp. 1348-1349, Dec. 2006.

[3] D. Neamen, Semiconductor Physics and Devices, New York, NY, McGraw Hill, 2003

[4] R. Baker, CMOS: Circuit Design, Layout, and Simulation, Rev. 2nd ed., Hoboken, NJ, John Wiley and Sons, Inc., 2008.

[5] Bradburn, S., "A high-voltage Buck converter relized with a low-voltage CMOS Process," University of Idaho, United States, 2010.

[6] Founds, J., Hess, H., Mentze, E., Buck, K., Richardson, M., "High-voltage series MOSFET output driver for nanometer technologies," Microelectronics Research and Communication Institute (MRCI), University of Idaho, Moscow, ID, unpublished.

[7] Buck, K. Li, H., Subramanian, S., Hess, H., Mojarradi, M., "Development and testing of high-voltage devices fabricated in standard CMOS and SOI technologies," Proc. 11th Annu. NASA Symp. VLSI Des., pp. 157-163, 2003.

[8] http://wiki.cs.uidaho.edu/index.php/Integrated_Passive_Components

POSTER SESSIONS

978-1-4577-1735-2/12 $31.00 © 2012 IEEE

PMOS Device Performance Improvement using Buried Contact Implants

S. Qin, T. McDaniel, L. J. Liu, R. Burke, Y. J. Hu, and A. McTeer
Micron Technology, Inc.
Boise, ID, USA

Abstract – An ultra-low-energy, high-dose, B-based implant was inserted after source/drain formation and before metal silicide contact deposition for PMOS devices. B beam-line implant and plasma doping (PLAD) using either B_2H_6 or BF_3 gases were utilized for this implant process. The resulting PMOS device performance showed significant improvements, including ~70 percent lower contact resistances, similar threshold and sub-threshold characteristics, and ~15 percent higher drive currents without degrading off current. PLAD is preferred on this application because of its much higher throughput in this process regime.

Deep Trench Patterning and Lift-off Resist in Micro-fluidic Devices

B. Pun, M. Mitkova, and P. Miranda
Boise State University
Boise, ID, USA

R. Zoller and M. Seibert
pSiFlow Technology Inc.
Boise, ID, USA

Abstract – This poster discusses the compatibility of silicon technology with micro-fluidic devices and its process flow. Cryogenic etch process is performed to create deep trench patterns for the formation of channels in the device. A bi-layer lift off technique is used to overcome challenges in creating thin aluminum traces.

Nano-ionic Conductive Bridge Memristors based on Chalcogenide Glasses: Electrical Performance Characterization and Modeling

M. R. Latif and M. Mitkova
Boise State University
Boise, ID, USA

Abstract – We report here the results on the fabrication of conductive bridge memory devices (CBM) based on Ge33Se67 doped with silver solid electrolyte. Silver is introduced in the chalcogenide matrix through photodiffusion. The conductive bridge memory devices are characterized by their I-V characteristics. It is shown that different compliance current contributes to achievement of different resistance states of the devices and multi-level switching. Modeling of the devices using SPICE software demonstrates that they can be considered as typical memristive memory.

Dependence of the Structure on Performance of Chalcogenide Glass Based Radiation Sensors

M. S. Ailavajhala, P. Chen, M. Mitkova, and Darryl P. Butt
Boise State University
Boise, ID, USA

D. Olesky, Y. G. Velo and H. Barnaby
Arizona State University
Tempe, AZ, USA

Abstract – Radiation detecting sensors are a vital area for research, where thin film resistance change based radiation sensors show great promise as the next generation of sensors. We propose two of such types of devices based on silver doped Chalcogenide Glasses. The resistance of the glass varies depending on the amount of diffused silver as a result of radiation. In this work we show two distinct device structures with different locations for the silver sources, and with the aid of simulations combined with experimental data, we show whether silver diffusing laterally or vertically throughout the structure is an optimal solution.

Flexible Photovoltaics

P. Salvador, M. Ostyn, and S. Parke
Northwest Nazarene University
Nampa, ID, USA

Abstract – Flexible solar cells offer many advantages over traditional rigid solar cells, including conformal surface applications and low-cost, roll-to-roll manufacturing processes. However, it is difficult to achieve adequate efficiency with these new flexible photovoltaics. This paper provides the background for this new NNU research project. A summary of existing flexible photovoltaic materials is given, including ultra-thin crystalline silicon, polycrystalline silicon, amorphous silicon, Copper-Indium-Gallium-Selenium (CIGS), and organics.

Measurement and Simulation of Hysteresis in Low Temperature Co-Fired Ceramic (LTCC) Electron Hop Funnel Characteristics

T. Rowe, M. Pearlman, and J. Browning
Boise State University
Boise, ID, USA

Abstract – Vacuum electron hop funnels have been constructed of Low Temperature Co-fired Ceramic (LTCC) to concentrate electron emission via electron "hopping" transport from a field emission array (FEA) electron source. The funnels are used to enhance emission uniformity and spatial distribution. Electrons are emitted up into the funnel, and current is sustained along the funnel wall by electron hopping transport. Electrons then exit the funnel in a denser, more uniform beam.

Gap in pagination due to unavailable paper.

Pages 45-46

Author Index

A
Abdelnaby, H. 5

B
Baker, R. J. 20, 33
Balagopal, S. 28
Barlow, F. 5
Benvenuti, A. 12
Bergman, K. xv
Betto, D. 12

C
Carnevale, G. 12

E
Elshabin, A. 5

F
Fantini, P. 12

G
Gallagher, W. J. xvi
Groothuis, S. K. 5

H
Hess, H. 37

J
Jindal, A. 1
Johnson, J. B. xiii

K
Kawamura, N. 9
King Liu, T. J. xiv
Koppula, R. 28

L
Lam, C. H. xii

M
Matsui, H. 9
McDaniel, I. 1

M
Mohan, R. 28
Moll, A. J. 20

N
Ng, J. H. 1
Nguyen, R. 16

P
Page, S. 37
Parker, R. S. 5
Phucas, N. 16
Potirniche, G. P. 5

S
Sarpeshkar, R. xi
Saxena, V. 20, 28
Snow, J. 9
Steneck, T. 16

T
Teo, B. B. 16
Tiwari, C. 16

V
VanAckern, G. 20
Ventrice, D. 12

W
Wajda, A. 37
Wei, W 1
Wong, S. N. 24

Y
Yap, K. 33

9781457717352